Plant pathology in agriculture

Plant pathology in agriculture

DAVID W. PARRY

Harper Adams Agricultural College, Newport, Shropshire

*The right of the
University of Cambridge
to print and sell
all manner of books
was granted by
Henry VIII in 1534.
The University has printed
and published continuously
since 1584.*

CAMBRIDGE UNIVERSITY PRESS
Cambridge
New York   Port Chester
Melbourne   Sydney

Published by the Press Syndicate of the University of Cambridge
The Pitt Building, Trumpington Street, Cambridge CB2 1RP
40 West 20th Street, New York, NY 10011, USA
10 Stamford Road, Oakleigh, Melbourne 3166, Australia

First published 1990

Printed in Great Britain at the University Press, Cambridge

British Library cataloguing in publication data
Parry, David W.
  Plant pathology in agriculture.
  1. Crops. Diseases
  I. Title
  632

Library of Congress cataloguing in publication data available

ISBN 0 521 36351 9 hard covers
ISBN 0 521 36890 1 paperback

Every effort has been made to describe correctly the uses of
agrochemicals mentioned in the text of this book. Users of agrochemicals
must read manufacturers' instructions on the labels of formulated
products, and should be aware of the approved uses of these products
within the UK under the Control of Pesticides Regulations, 1986 (as part
of the Food and Environmental Protection Act, 1985). Neither the
author nor the publisher will accept any responsibility for any damage,
accidental or otherwise, arising from the use of any of the agrochemicals
mentioned in this book.

Back cover: photograph courtesy of ADAS Aerial Photography Unit, MAFF,
Cambridge, Crown Copyright reserved. Spread of *Phytophthora infestans* from
foci of infection with the prevailing wind.

WV

To Carol, Jonathan and Andrew

Contents

Contents

List of figures

Preface

This book is aimed at students, consultants, advisers, farmers, the agrochemical industry and any other parties interested in aspects of plant pathology applied to agriculture. It is hoped that the text will be sufficiently clear for use by students of agriculture at all levels of higher education, and sufficiently comprehensive to provide a basis for undergraduate and postgraduate studies in plant pathology as part of crop protection.

The book is divided into two main sections, the first on the principles and the second on the practice of plant pathology in agriculture. In the first section, five basic questions are posed: what is disease? what causes disease? how does disease build up? what effect does disease have on the crop? and, how do we control disease? Attempts are made to answer the questions by outlining the principles of plant pathology giving brief examples from agriculture where applicable. The second section puts these principles into practice and describes details of the biology, recognition and control of specific diseases of arable crops.

Acknowledgements

I am indebted to the following people for help in the preparation of this book: R. C. Anslow, M. J. C. Asher, K. W. Bailiss, R. A. Bayles, J. D. S. Clarkson, L. J. Cocke, R. J. Cook, J. W. Deacon, W. E. Downer, N. J. Giltrap, P. Gladders, D. Habeshaw, N. V. Hardwick, P. Payne, R. W. Polley, D. J. Royle, R. C. Shattock, M. C. Shephard, M. C. Shurtleff, J. B. Sinclair, B. E. J. Wheeler and D. J. Yarham. Particular thanks are due to J. T. Fletcher and T. F. Preece for their overview of the script. I also thank my typist at Harper Adams, Mrs C. Y. J. Cochrane. Acknowledgement to those who kindly supplied photographs is made in the legends to individual plates. Lastly, I thank my patient wife Carol for her help in editing and checking the manuscript.

I PRINCIPLES

1 What is disease?

Introduction

Disease has been defined in various ways, for example:

'Morbid condition of body, plant, or some part of them, illness, sickness' (Concise Oxford Dictionary).

Talboys *et al.* (1973) defined plant disease as

'A harmful deviation from normal functioning of physiological processes'

whereas Agrios (1978) defined plant disease as

'Any disturbance brought about by a pathogen (organism which causes disease) or an environmental factor which interferes with manufacture, translocation, or utilization of food, mineral nutrients, and water in such a way that the affected plant changes in appearance and/or yields less than a normal, healthy plant of the same variety'.

Such a comprehensive definition as that of Agrios is a reflection of the complex nature of disease. Indeed Tarr (1972) considered that there is no entirely satisfactory definition of disease. Part of the problem in defining disease is that of reference to the 'normal' plant. Natural variations in plant populations together with subtle, almost undetectable alterations in metabolism caused by some pathogens may make it difficult to find a 'normal', disease-free plant. In addition, each plant in a productive crop is already suffering extreme competition for light, nutrients and water with neighbouring plants, which inevitably results in stress and affects metabolism. In spite of these difficulties plant pathologists' understanding of disease assumes that the plant is in some way abnormal and that some external factor is the cause of this. Under the very broad definition as proposed by Agrios, disease could be caused by any one or a combination of the following:

Nutrient deficiencies or excesses
Toxic materials in the soil or atmosphere
Infestation by pests
Colonisation by flowering plants or algal parasites
Infection by micro-organisms (fungi, bacteria, mycoplasmas)
    and viruses

Plant pathologists in many countries study all of these causes of disease. However, in the UK most plant pathologists confine themselves to studies of infection by micro-organisms and viruses, and this book will not attempt to break with the UK tradition.

Plant diseases can have a considerable impact on the general public with respect to food supply, and also safety, quality and diversity of foods available. There are no up-to-date estimates of world losses attributable to plant disease, but Agrios (1988a) suggests that crops worth over $9 billion are lost to diseases each year in the USA. The ever increasing world population requires higher productivity on a constantly declining area of agricultural land. A better understanding of plant diseases, together with the development of more effective disease control measures, will play a vital role in providing the world with an adequate and varied supply of safe, high quality food.

Modes of nutrition of micro-organisms

In order to understand how infection by micro-organisms can cause disease, one must first understand a little about their modes of nutrition. Most micro-organisms are beneficial to man in that they break down dead and decaying organic material. Such substances utilised by micro-organisms for growth are called **substrate**. Organisms which feed in this way are **saprophytes**. However, there are other micro-organisms which live on other living organisms, their **hosts**. They take nutrients from their hosts in a way that does not benefit the host at all. Such organisms are **parasites**. If there is mutual benefit in the relationship between micro-organism and host, for example in the case of some *Rhizobium* spp. forming root nodules on legumes, the relationship is **symbiotic**. The term parasite is often used interchangeably with **pathogen**. However, in the strictest sense, they are not the same. A pathogen is an organism which causes disease and is often, but not always, a parasite; for example, certain saprophytic fungi in the soil produce toxic substances which can damage plants. Some surface leaf-growing fungi, e.g. sooty moulds, grow on honeydew secreted by aphids, cover the photosynthetic area and reduce growth. In each example the pathogen is not feeding on the plant but nonetheless is causing disease.

In the relationship between the pathogen and the host, a further dimension relating to mode of nutrition needs explanation. Traditionally, parasites have been divided into **obligate** and **facultative parasites**. An obligate parasite can live and reproduce only by feeding on living plant material. Obligate parasites are often referred to as **biotrophs**. In practice it is difficult or impossible to grow obligate parasites such as *Puccinia striiformis* (a fungus causing yellow rust of wheat) in the laboratory on agar or other media. In contrast, facultative parasites can feed equally well on dead or decaying material, or living plants. Facultative parasites cause problems particularly if a host is weakened in any way. For example, soil-inhabiting fungi such as *Pythium* spp. cause damping-off of seedlings of many genera, especially if the soil is waterlogged. Facultative parasites may be **necrotrophs**, i.e. organisms which kill host tissues and then utilise the dead host tissues as a substrate.

Symptoms of disease
Growth of pathogens induces changes in host plants; thus disease is manifested as symptoms. Disease symptoms may be categorised as discolorations, abnormal growth, rots and physiological wilts. Infection may also result in subtle, but important, alterations in the physiology of the plant. Disease symptoms are discussed further in Chapter 4.

The diagnosis of some diseases is based on the appearance of some stage in the life history of the pathogen itself. For example, rust diseases are so named because the spores of the pathogen are often rust-coloured. Sometimes, the fungal presence may not be so obvious; for example, small black dots in a lesion caused by the fungus *Septoria tritici* on wheat are fruiting bodies called pycnidia. Such structures are just visible to the naked eye and distinguish *Septoria tritici* from *S. nodorum* which has almost invisible pycnidia. Often the outward, visible symptoms of disease need to be followed up with confirmatory laboratory isolation of the pathogen, or microscope work. Some bacterial and virus diseases are very difficult to diagnose in the field and a laboratory test is essential.

Some disease symptoms are localised and obviously associated with the immediate presence of a pathogen, e.g. *Septoria* lesions on the leaf of a wheat plant. Other symptoms may be manifested further away from the pathogen, and an association between pathogen and symptoms must be established to prove pathogenicity, e.g. vascular wilt diseases. Another problem with the association between symptoms and pathogens is presented by the many and various leaf spots found on almost all arable crops. Such lesions may appear not as a result of pathogen invasion, but as a result of adverse weather conditions, agrochemical spray damage, fertiliser scorch, physiological disorder, pest damage or nutrient deficiencies. In such situations, some proof of the pathogenicity of a suspect organism is required. This proof of pathogenicity is one of the fundamental principles of plant pathology and the need for it was recognised by Koch in 1884, at the start of the science of microbiology.

Koch's postulates

1. The organism must be consistently associated with the symptoms of the disease.
2. The organism must be isolated and grown in pure culture.
3. The organism must be inoculated onto healthy hosts of the same species from which it was originally isolated, and must reproduce the same symptoms as originally observed.
4. The organism must be re-isolated and have the same characteristics as the original isolate.

Although these principles are over 100 years old and have a number of limitations, they are still useful in proving the pathogenicity of plant pathogens. Experimental problems can arise in attempts to prove pathogenicity using Koch's postulates. Some symptoms are physically separated from the pathogen, as already mentioned with vascular wilt diseases. In addition, symptoms may be a result of simultaneous infection by several pathogens. For example, soft rot of potatoes in store is most frequently attributable to the bacterium *Erwinia carotovora* subsp. *carotovora*, but other bacteria have also been implicated. Stem base rot of cereals can be a result of infection by the fungi *Pseudocercosporella herpotrichoides* (the cause of eyespot), *Fusarium* spp. (the cause of brown foot rot), *Rhizoctonia cerealis* (the cause of sharp eyespot) or any combination of these pathogens. Fungi often act synergistically, giving rise to more damage together than the sum of their individual effects. A further potential problem concerns inoculation with pathogens into healthy host plants. This requires the right type of inoculum. In fungi, inoculum usually consists of spores which may have to be induced in culture by the use of special growth media and/or environmental conditions, for example near ultra-violet (black) light. Bacterial cells can also be multiplied in culture but viruses can multiply only in living plant tissue. The inoculum then needs to be applied in a way which simulates nature. Rain-splashed fungal spores need to be sprayed over plants. Most bacterial plant pathogens require wounding at inoculation and particular problems arise with many viruses, which can only be transmitted by specific vectors such as aphids. Finally, environmental conditions have to be created which are optimal for infection and symptom development. Perhaps the main limitation in the use of Koch's postulates concerns the requirement for the organism to be isolated and grown in pure culture. Many important pathogenic fungi will not grow on artificial growth media, e.g. *Erysiphe graminis* which causes powdery mildew of cereals. Obligate parasites, such as viruses which can only multiply in living plants, pose obvious problems here.

It is clear that much experimentation and a full understanding of the biology of the pathogen are required before Koch's postulates can be satisfied in plant pathology. Indeed the problems associated with producing infected plants artificially may suggest that it is almost impossible for disease to develop in the field. Obviously this cannot be the case. However, it is evident that there are several critical factors which influence disease development.

Factors affecting disease development

Disease is a complex interaction of factors associated with the pathogen, host and environment (Figure 1.1). The practice of plant pathology in agriculture demands a full understanding of the influence of the host, pathogen and environment on disease development. Indeed, disease will not build up unless there is an active pathogen, a susceptible host and suitable environmental conditions for infection, colonisation and reproduction of the pathogen.

Characteristics of the pathogen important in disease development will be dealt with first, followed by influences of the

Figure 1.1. The interaction of factors involved in the development of disease in plants, often known as the 'disease triangle'.

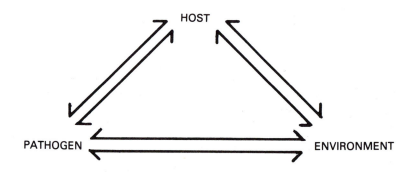

environment on the host and pathogen and vice versa. The host/pathogen relationship is considered in detail in Chapter 5 under disease resistance.

Pathogen

In order for disease to be initiated some infective propagules (inoculum) of the pathogen, e.g. fungal spores, bacterial cells, must be present. To some extent, the amount of disease which builds up during the season is affected by the quantity of initial inoculum. With some pathogens, however, the effect is transient because once the first infections are established, the pathogen produces more inoculum on the affected plants. This leads to a spread of the pathogen and development of disease on other plants throughout the crop, e.g. powdery mildew disease of many crops. Other pathogens infect plants, but do not contribute further inoculum until the crop matures or the affected plant dies; in such cases development of disease depends almost entirely on the amount and distribution of the initial inoculum. For example, if there is a large population of sclerotial inoculum, a susceptible variety and optimum conditions for infection, 60% of a crop of oilseed rape plants may become infected with the fungus *Sclerotinia sclerotiorum* (the cause of stem rot). However, a small localised focus of sclerotia would restrict the disease to a small proportion of the crop.

The ability of a pathogen to infect and colonise plant tissue is another factor which influences disease development. The balance between the host plant's ability to restrict disease and the pathogen's ability to cause disease is critical. The ability to cause disease depends in some cases on the genetic characteristics of the pathogen. Aggressive qualities of the pathogen, such as the ability to grow quickly from its source, and to penetrate and colonise host tissue rapidly, may also be important. Finally, the ability of the pathogen to compete with other nearby micro-organisms may be highly significant in determining whether or not disease occurs. This is particularly important in soil-borne pathogens. Pathogens which compete well in soil for dead and decaying material, e.g. the fungus *Fusarium culmorum*, will be much more durable in soil than pathogens such as *Gaeumannomyces graminis* (causing take-all of cereals) which find the soil a difficult place to survive in because of competition from other soil saprophytes. Even if rotations could be made with crops not susceptible to *F. culmorum*, it would be impossible to eradicate the fungus totally because of its ability to live as a saprophyte in the soil. However, a break from cereals will reduce take-all to a negligible level: *G. graminis* is simply starved out when its wheat straw substrate rots away.

Environment

It is evident that almost all diseases occur only, or develop best, during the warmer parts of the year. It is also widely recognised that diseases appear after periods of high humidity or rainfall. These general observations indicate that environmental conditions can greatly affect disease development. The main factors to be considered are temperature, moisture and soil.

Temperature

Certain minimum temperatures are required before plants and pathogens can grow and reproduce. Temperatures in winter in the UK normally inhibit the growth of both, indeed extremely cold winters may kill plant tissue and reduce pathogen populations. For example, early-sown winter barley often becomes infected by the fungus *Erysiphe graminis* which causes powdery mildew. If leaves die as a result of frost attack during winter, this biotrophic fungus also dies and inoculum levels are much reduced for the subsequent infections which occur routinely in the spring. The optimum temperature requirement for disease development varies considerably, even within one genus of fungi. For example, snow mould of cereals caused by the fungus *Fusarium nivale* occurs under snow-cool damp conditions at temperatures of 5 °C or less. In contrast, other species of *Fusarium* which cause ear blight of cereals (*F. culmorum*, *F. avenaceum*, *F. graminearum*) are only a problem in summer at temperatures of *c.* 20 °C and when humidity is very high. In general terms, however, higher temperatures are more conducive to disease development.

As Table 1.1 shows, the time taken for the rust fungus *Puccinia graminis* to germinate on a plant surface, penetrate and colonise plant tissue and produce fresh pustules (**latent period**) rapidly decreases with increase in temperature. Epidemics of black stem rust are most likely to occur during summer at a time when fresh infections can occur at 5-day intervals. The effect of temperature on

Table 1.1. *The effect of temperature on the latent period of* Puccinia graminis *(causing black stem rust) on wheat*

Temperature (°C)	*c.*	0	4.5	10.5	12.5	19	21	24
Latent period (days)		85	22	15	12	9	7	5

(After Stakman & Harrar, 1957.)

development of bacterial diseases can be equally striking. The bacterium *Erwinia carotovora* subsp. *carotovora*, the cause of soft rot of potato tubers, is most active and causes most losses in unventilated potato stores where temperatures can reach 20 °C. The effect of temperature on virus diseases is more complex. Temperature influences the growth of the plant, multiplication of the virus and, in addition, the vector.

Moisture

Moisture exists as rain or irrigation water on the plant surface or in the soil, and as relative humidity (R.H.) in the air or as dew. In these forms moisture can influence disease development in many ways and may be the critical factor which determines disease occurrence and development. For example, potato blight, caused by the fungus *Phytophthora infestans* will develop only at high relative humidities and is inhibited if R.H. drops below 90%. Consequently the disease is most frequently a problem during wet summers and in areas of high rainfall. In fungi, the most important influence of moisture seems to be upon the germination of fungal spores on the host surface and the penetration of the host by the germ-tube. Most fungal spores require free water or very high relative humidities for infection. The exception to this rule occurs with the powdery mildews, where spore germination is adversely affected by free water; some mildews, e.g. *Erysiphe pisi*, can germinate at low R.H. Once fungi have penetrated and are established in plants, the presence of free water or very high humidities is less important. However, sporulation may be stimulated by periods of high relative humidity.

In many diseases affecting plant roots, the severity of the disease is proportional to the soil moisture content, e.g. *Pythium* spp. causing damping-off in many crops are most severe in wet, poorly drained soils. *Pythium* spreads in the soil primarily by motile zoospores which swim in the soil moisture. There may be a further effect of excess soil moisture on disease development. Plants in waterlogged soil are under stress as a consequence of reduced oxygen and lower temperatures around the roots. Stressed plants are more prone to attack by facultative parasites such as *Pythium* spp. Potato growers may face a dilemma with regard to the effect of moisture on disease development. Periods of dry weather during tuber initiation (June and July) favour the development of common scab on potatoes caused by the bacterium *Streptomyces scabies*, whereas wet weather favours the blight fungus *Phytophthora infestans* and the canker phase of powdery scab caused by the fungus

Spongospora subterranea. The decision to irrigate potato crops during early summer is clearly a difficult one to make as regards disease development. However, the yield benefits of irrigation often outweigh possible yield losses as a result of disease.

Moisture has a marked effect on bacterial disease development. Bacterial pathogens are usually dispersed by water and multiply more rapidly in wet conditions.

Soil

The main factors in the soil, other than moisture, which may influence disease are soil pH and nutrient status. The effect of soil pH on the disease clubroot of brassicas caused by the fungus *Plasmodiophora brassicae* is dramatic. At low pH (5.7 or below), clubroot can be severe; however, its severity drops between pH 5.7 and 6.2 and the disease is completely inhibited at pH 7.8. In contrast, common scab of potatoes (caused by the bacterium *Streptomyces scabies*) can be severe above pH 5.2, but in more acid soils its incidence is low. Manipulation of pH by the farmer in order to reduce disease may be a good idea in theory but in practice it is more important to establish the correct soil pH for the crop. However, liming of soil, to increase pH and thus reduce clubroot, is a fairly widespread practice.

Soil nutrient status, particularly the amount of nitrogen fertiliser, can also affect disease development. Mildew of cereals caused by the fungus *Erysiphe graminis* is more severe in soils over-fertilised with nitrogen. Under such conditions, the cereal plants produce many shoots very quickly. Such a soft dense growth of tissue, with its consequent increased localised humidity is very susceptible to mildew attack. Take-all of cereals (caused by the fungus *Gaeumannomyces graminis*) is more severe in soils deficient in nitrogen fertiliser. Plants grown in such soils tend to be weaker, and have less well developed root systems. They are consequently less able to withstand attacks by the fungus. A well balanced fertiliser programme, which corrects any nutrient deficiencies, will result in more take-all tolerant plants.

The less obvious effects of the host on the environment in relation to disease development will now be considered briefly.

Populations of potential host crop plants often have important effects on their surrounding microclimate which can dramatically influence disease development. For example, late blight of potatoes caused by the fungus *Phytophthora infestans* becomes troublesome in more mature potato crops when crop canopies become dense.

Under dense canopies, relative humidities increase which allows *P. infestans* to flourish. Fungicide spray programmes aimed at controlling potato blight normally start when leaves of individual plants meet along the row of potatoes.

Further reading

Agrios, G. N. (1988). *Plant Pathology,* 3rd edn. New York: Academic Press.

Gareth Jones, D. (1987). *Plant Pathology: Principles and Practice.* Milton Keynes: Open University Press.

Johnston, A. & Booth, C. (ed.) (1983). *Plant Pathologist's Pocketbook.* Slough: Commonwealth Agricultural Bureaux.

Manners, J. G. (1982). *Principles of Plant Pathology.* Cambridge: Cambridge University Press.

Tarr, S. A. J. (1972). *Principles of Plant Pathology.* London: Macmillan.

2 What causes disease?

Fungi as plant pathogens

General characteristics

Fungi are the most important plant pathogens. Plant pathogenic fungi are microscopic organisms which, unlike plants, lack chlorophyll and conductive tissue. Of the 100 000 or more species of fungi so far described, about 50 cause diseases in Man (most often superficial skin disorders such as athlete's foot) but more than 8000 species are known to cause disease in plants. A typical fungus consists of a vegetative body (**mycelium**) made up of individual branches (**hyphae**) which may or may not have cross-walls (**septa**). In the primitive or lower fungi, the mycelium consists of a continuous mass of cytoplasm and nuclei unbroken by septa (**coenocytic**). In higher fungi, the mycelium is divided by septa into units which each contain one or a few nuclei. Some lower fungi lack a true mycelium and consist of a naked amoeboid multinucleate plasmodium (*Myxomycetes* or slime-moulds).

Reproduction

Fungi reproduce chiefly by means of spores. Spores are specialised propagative or reproductive bodies, consisting of one or a few cells, which are formed sexually or by an asexual process. The enormous variety of size, shape and methods by which fungi produce spores is used as the basis for their classification. Asexual spores are produced in sporangia which develop in lower fungi on specialised hyphal branches called sporangiophores. Sporangia can germinate directly by growth of a germ tube or they can release a number of motile, flagellate zoospores. In higher fungi, asexual spores called conidia are formed from specialised hyphae, conidiophores. These conidia may be produced in flask-shaped structures. These include conidia, which may be formed simply on hyphae in exposed pustules as in powdery mildew, or pycnidia as in *Septoria* spp. Other types of asexual spores include specialised resting spores called chlamydospores. These are thick-walled spores produced at the ends of hyphae or by conversion of intercalary cells. Some spores also have the ability to change from thin-walled conidia to thick-walled chlamydospores, for example macroconidia of *Fusarium culmorum*.

Sexual reproduction is a feature of most plant pathogenic fungi, and the process by which this occurs is important in classification. In most fungi, both male and female sexually active cells (gametes) are produced on the same mycelium. If fertilisation can occur within the same mycelium, the fungus is termed homothallic. If male gametes can fertilise only female gametes of another sexually compatible mycelium, the fungus is termed heterothallic. Sexual reproduction can give rise to resistant overwintering bodies such as oospores, or short-lived spores such as ascospores. Most fungi can reproduce by either a sexual or an asexual process, although the sexual stage tends to be less obvious. Asexual spores are usually responsible for disease epidemics and are therefore more often encountered by plant pathologists. Fungi which appear to have no sexual or perfect stage have been grouped artificially into the Fungi Imperfecti or Deuteromycotina.

Taxonomy

Numerous classification schemes have been proposed for fungi (see Hawksworth, Sutton & Ainsworth, 1983) and differences of opinion still exist as to the precise taxonomic position of certain species. However, one of the most widely adopted systems, published by Johnston & Booth (1983) will be used. Divisions end in '-mycota'; Sub-divisions in '-mycotina'; Classes in '-mycetes'; Sub-classes in '-mycetidae'; Orders in '-ales'; Families in '-aceae'. For example, the potato blight fungus *Phytophthora infestans* is classified as follows:

Division	Eu*mycota*
Sub-division	Mastigo*mycotina*
Class	Oo*mycetes*
Sub-class	Peronosporo*mycetidae*
Order	Peronospor*ales*
Family	Peronospor*aceae*
Genus	*Phytophthora*
Species	*infestans*

Most fungi are not plant pathogens and the following classification is an aid to understanding how plant pathogenic fungi fit into the large groups of non-pathogenic fungi. A brief description is given of Classes within Sub-divisions which include plant pathogenic species. Representative spore types are also shown in Figure 2.1.

I MYXOMYCOTA (slime moulds)

Plasmodium (aggregate of naked fungal cells) formed in host. Plasmodiophoromycetes contains Plasmodiophorales, motile zoospores produced, e.g. *Plasmodiophora brassicae* (the cause of club root of brassicas).

II EUMYCOTA (true fungi)

(1) Mastigomycotina
Motile zoospores produced.

(a) Chytridiomycetes No true mycelium, asexual and sexual reproduction by motile zoospores. Includes Chytridiales, e.g. *Synchytrium endobioticum* causing wart disease of potato tubers.

(b) Oomycetes Aseptate mycelium produced, asexual reproduction by motile zoospores or sporangia. Oospores formed by sexual reproduction. Includes Peronosporales (downy mildews) and *Phytophthora infestans*, the cause of late blight of potatoes.

(2) Zygomycotina
Aseptate mycelium produced, asexual reproduction by non-motile spores, zygospore formed by sexual reproduction.

(a) Zygomycetes Includes Mucorales, e.g. *Rhizopus* spp. which cause soft rot of fruits.

(3) Ascomycotina (Ascomycetes)
Septate mycelium produced, asexual conidia and sexual ascospores formed in a variety of ways.

(a) Hemiascomycetes Asci produced singly and unprotected. Includes Taphrinales, e.g. *Taphrina deformans* (the cause of peach leaf curl).

(b) Plectomycetes Ascocarps (contain asci containing ascospores) produced in a closed cleistothecium. Includes Erysiphales, e.g. powdery mildews.

(c) Pyrenomycetes Asci produced in flask-shaped perithecium with an apical hole. Includes Hypocreales (light, soft perithecium), e.g. *Claviceps purpurea* the cause of ergot of cereals and Sphaeriales (dark, hard perithecium), e.g. *Gaeumannomyces graminis* causing take-all of cereals.

(d) Discomycetes Asci produced in cup-shaped apothecium. Includes Helotiales, e.g. *Sclerotinia sclerotiorum* (the cause of stem rots).

(e) Loculoascomycetes (bitunicate ascomycetes) Asci produced in aggregation of vegetative hyphae (pseudothecium). Includes Dothideales, e.g. *Mycosphaerella brassicicola* which causes ring spot of brassicas.

(4) Basidiomycotina (Basidiomycetes)
Asexual conidia e.g. uredospores, and sexual basidiospores formed. Sclerotia and rhizomorphs (hyphal aggregations) may also be formed.

(a) Hemibasidiomycetes Basidiospores produced on septate basidia, no basidiocarps formed. Includes Uredinales (rust fungi), and Ustilaginales (smut fungi).

(b) Hymenomycetes Basidiospores produced on aseptate basidia, complex fruiting bodies (basidiocarps) formed. Includes Agaricales, e.g. *Armillaria mellea* (the cause of root rot of trees).

(5) Deuteromycotina (Fungi Imperfecti)*
Asexual classification.

(a) Coelomycetes Conidia borne in specialised structures. Includes Melanconiales, conidia formed on cushion-like structures, acervuli, e.g. *Colletotrichum* spp. (anthracnose), and Sphaeropsidales,

---

* Fungi Imperfecti have often been discovered to have a sexual stage which leads to re-classification, frequently as Ascomycetes. This leads to confusion in nomenclature when a fungus is classified with two names, e.g. *Septoria nodorum* (Coelomycete)=*Leptosphaeria nodorum* (Loculoascomycete).

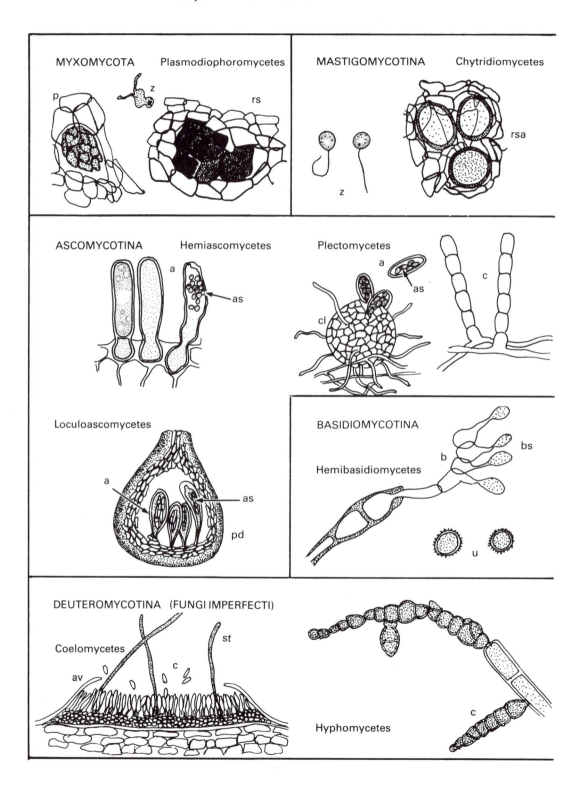

MYXOMYCOTA   Plasmodiophoromycetes

MASTIGOMYCOTINA   Chytridiomycetes

ASCOMYCOTINA   Hemiascomycetes

Plectomycetes

Loculoascomycetes

BASIDIOMYCOTINA

Hemibasidiomycetes

DEUTEROMYCOTINA (FUNGI IMPERFECTI)

Coelomycetes

Hyphomycetes

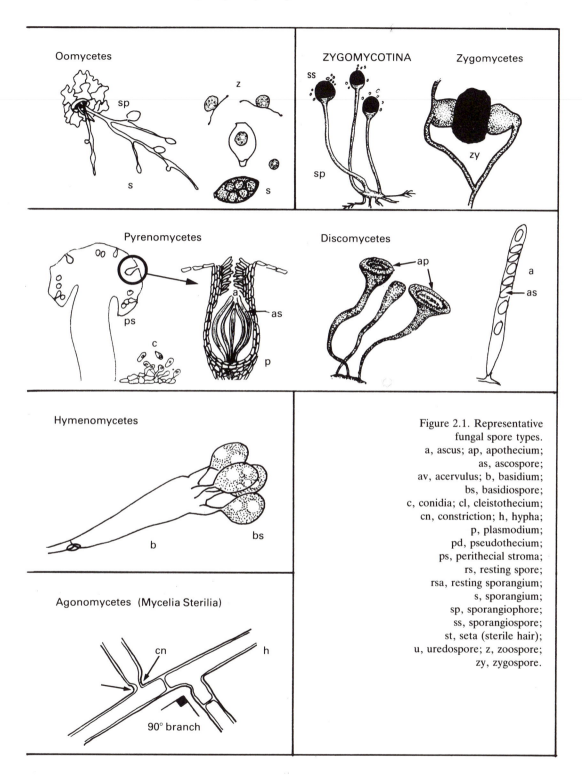

Figure 2.1. Representative
fungal spore types.
a, ascus; ap, apothecium;
as, ascospore;
av, acervulus; b, basidium;
bs, basidiospore;
c, conidia; cl, cleistothecium;
cn, constriction; h, hypha;
p, plasmodium;
pd, pseudothecium;
ps, perithecial stroma;
rs, resting spore;
rsa, resting sporangium;
s, sporangium;
sp, sporangiophore;
ss, sporangiospore;
st, seta (sterile hair);
u, uredospore; z, zoospore;
zy, zygospore.

conidia formed in flask-shaped pycnidia, e.g. *Septoria* spp. (causing leaf spots).

(b) Hyphomycetes Conidia not borne in specialised structures. Includes Hyphales (Moniliales), e.g. *Alternaria* spp. (the cause of leaf spots).

(c) Agonomycetes (Mycelia Sterilia) No common spore forms. Includes Agonomycetales, e.g. *Rhizoctonia cerealis* which causes sharp eyespot of cereals.

Dispersal

The dispersal of spores of plant pathogenic fungi is of prime importance in the spread of disease. It is convenient to separate fungal dispersal into three categories: autonomous; insect and other animal; air and water. Many species of fungi use more than one method of dispersal and the combination of air and water is very common.

Autonomous

Autonomous dispersal of fungi is simply dispersal in normal crop husbandry. Farmers plough fields and, in doing so, may spread soil-borne plant pathogens. Hence the increasingly important virus disease of barley, barley yellow mosaic virus (BaYMV), which has the soil-borne fungal vector *Polymyxa graminis*, tends to spread along cultivation lines in infested fields. It may be spread to uninfested fields by contaminated soil on machinery or boots. Other routine farming practices such as irrigation may also act as dispersal agents for pathogens, for example, potato blight spores (*Phytophthora infestans*) are very effectively dispersed and, moreover, infection is encouraged, by intensive irrigation.

Insects and other animals

Many complex and diverse relationships exist in this category. There are several interesting examples of fungal dispersal by insects, including ergot disease of cereals. Following initial infections from spores produced by ergots (sclerotia) in the soil, the subsequent spread of the ergot fungus (*Claviceps purpurea*) within a cereal crop is effected by insects attracted to affected cereals by the honeydew in which conidia are produced. There are a few seemingly fortuitous cases of dispersal by insects. Dutch elm disease, caused by the fungus *Ceratocystis ulmi*, is spread by the European elm bark beetle (*Scolytus multistriatus*) and the native elm bark beetle (*Hylurgopinus rufipes*). The fungus overwinters in the bark of dead or dying elm trees and logs. Elm bark beetles lay their eggs between the bark and wood, and adult beetles emerge from diseased trees covered with sticky spores of *C. ulmi*. Beetles then burrow through the bark and into the wood of vigorous healthy trees on which they feed. Spores of the pathogen are deposited in the wounded tissue of the tree and dispersal of the fungus is thus effected. There is some evidence that aphids feeding on potatoes with potato blight can become contaminated with sporangia and transmit the fungus over short distances. Some insects play important roles in the life cycles of certain pathogens. For example, insects can bring about the transfer of spores of rust fungi from one mating type to another, which is essential for the development of other stages of these fungi.

There are some reports of root-infecting pathogens entering wounds made by nematodes. Cotton wilt caused by *Fusarium oxysporum* f.sp. *vasinfectum* appears to be associated with damage caused by root knot nematodes.

It is clear that any animal which moves through a diseased crop into a healthy crop is a potential dispersal agent for fungal diseases. Fungal spores can very easily become attached to the fur of mammals and the feathers of birds. Some spores, particularly resting spores, will, as do many plant seeds, pass through the intestines of animals intact. The causal agent of powdery scab of potato, *Spongospora subterranea*, can pass through the intestines of cattle unharmed; hence it is wise not to feed cattle with diseased potatoes, as contaminated slurry may then be spread onto fields.

Man is sometimes particularly effective at fungal dispersal, and has been implicated in both long and short-distance spread of fungi. The vast international trade in plant material including seeds, horticultural stock and fruit may give rise to importation of fungal diseases if plant material is not inspected thoroughly. An example of long-distance dispersal in this way again concerns Dutch elm disease. The aggressive pathogen strain of *Ceratocystis ulmi* was most probably introduced in elm timber imported to Europe from North America. Tourists may bring infected plants or seeds into a country illegally. Man can also spread diseases relatively short distances, for example within and between fields. Many crops, such as potatoes and soft fruits, are vegetatively propagated. Such propagation may result in contamination of offspring particularly by systemic fungal pathogens, e.g. *Verticillium dahliae*, the cause of vascular wilt of numerous crops including potatoes and strawberries.

Air and water

Most important plant pathogens are dispersed in the air. Water is also often involved in air-borne dispersal of spores. However, the presence of water or a high relative humidity is a pre-requisite for many fungi before active sporulation and discharge into the air take place. In addition, the impaction of rain droplets on foliage may be important in launching air-dispersed spores.

In any form of air travel, there are three distinct phases; the take-off, the journey through the air and the landing. The take-off phase of fungal dispersal is interesting in that the fungal spores must cross a thin layer of still air, the boundary layer, which covers all plant and soil surfaces. Some fungi have evolved ingenious ways of crossing this layer by actively discharging their spores. Normally in active discharge, turgid cells are involved, so moist conditions are required. Many species of Ascomycotina can actively discharge sexually produced ascospores from the spore case, the ascus. The ascus can be considered as a water-squirting mechanism which, when it bursts, projects the ascospores up to 600 mm from the ascus. The swelling and bursting of the ascus can only occur during high relative humidity or rain and in many diseases caused by members of the Ascomycotina, including apple scab caused by *Venturia inaequalis*, there is a close relationship between periods of rain and the presence of ascospores in the air. Among the sub-division Basidiomycotina, there is also evidence of active discharge. The basidium in many species violently projects the basidiospore into the atmosphere. Generally the distance projected does not exceed 1 mm. The mechanism by which this occurs has been investigated at some length without any consensus of opinion emerging. A drop-like swelling often appears around the point of attachment of the spore; this may be involved in the discharge process. Although most fungi require an increase in relative humidity before active discharge occurs, there is a curious phenomenon in a number of downy mildew fungi including *Pseudoperonospora humuli* (the cause of downy mildew of hops) which actively discharge sporangia after a sudden decrease in R.H. The sporangiophores twirl rapidly, flicking off the finely attached sporangia.

Although active discharge of spores is an interesting subject to study, most plant pathogenic fungi have no known method of violent spore discharge and spores are liberated passively. For such fungi, external factors have to provide a means of crossing the boundary layer of still air. The most frequent means by which this is achieved is the action of wind and rain. Relatively low wind speeds have been found sufficient both to break the connection between

the conidia and the conidiophore, and to take the free spore through
the boundary layer. Wind speeds as low as 0.4 m/sec can allow
spores of cereal powdery mildew caused by *Erysiphe graminis* to
become air-borne. Spores of black stem rust of wheat (caused by
Puccinia graminis) require a little extra wind, 1.7 m/sec. Stronger
winds will result in an increase in numbers of spores liberated into
the air. Heavy rain impacting on aerial parts of the plant may also be
responsible for catapulting dry spores through the boundary layer.

Once spores are air-borne, effective plant pathogens need to be
dispersed at least locally within the host crop and preferably over
longer distances to new healthy crops. As distance increases from
the source or focus of infection, initially at least, incidence of
disease will decline. Distribution of a foliar disease within a crop,
for example potato blight, will follow a pattern close to the
prevailing wind (Figure 2.2). The pattern of spread may not be

Figure 2.2. Spread of
Phytophthora infestans,
the cause of late blight of
potatoes; from foci of
infection (darker patches)
with the prevailing wind.
(Black and white infra-red
film, ADAS Aerial
Photography Unit, MAFF,
Cambridge, Crown
Copyright reserved.)

circular from the original focus but may spread almost like a plume of smoke from a chimney. This is frequently the pattern of air-borne spread of plant pathogens over short distances within a host crop. However, spores of some plant pathogens may be taken up much higher into the atmosphere on thermals and other air currents. There is then potential for spores to travel many miles on intercontinental winds. Hirst, Stedman & Hurst (1967) sampled the atmosphere for spores for a single day over the North Sea, using an aircraft. They identified a cloud of the facultative parasite *Cladosporium* at 1000 m which was over 150 km in length. More significant plant pathogens such as rusts may also be able to travel long distances in air currents. Outbreaks of black stem rust of wheat (caused by *Puccinia graminis*) have occasionally been reported in the south-west of England. The spores must have been blown in from the Mediterranean areas of Europe or North Africa where the disease is quite common. There is also evidence for the fungus spreading over long distances in the USA. It is generally believed that the Atlantic is a barrier to the spread of air-borne pathogens, reflecting the relatively harsh conditions which spores have to endure in high-altitude, long-distance dispersal. The air becomes progressively colder at higher altitude; this could actually prolong the life of a spore if it was subjected to a progressive temperature drop without large fluctuations. The quantity of ultra-violet light in the atmosphere progressively increases with altitude. This could be a problem for spores, which may be killed by high levels of UV radiation. But perhaps the most significant problem for spores to overcome is that of desiccation. The constant drying effect of the wind will, given time, kill all air-borne spores. However, despite these problems, uredospores of some rust species are able to survive for several weeks high in the atmosphere.

Given that spores have been effectively launched and dispersed, at least a short distance in the air, they have the problem of landing. Just as spores have a problem getting air-borne because of the boundary layer of still air, they must also penetrate this layer in order to land on host plant material. Gravity plays a part in the deposition of spores when the air is still and free from turbulence; wind and rain are also of importance.

Size and shape of the fungal spores play a large part in their ability to be effectively deposited. Air-borne fungal spores are mostly spherical or ovoid with diameters of 5–20 μm. In still air, large spores, such as those of the cereal powdery mildew fungus *Erysiphe graminis* have been found to fall faster than smaller spores, giving them a higher terminal velocity. The higher the terminal velocity, the greater the chance of penetration of the boundary layer. Hence it has been proposed that the relatively large conidia of leaf and stem pathogens such as rusts and mildews are adapted to penetrate boundary layers on comparatively small areas of tissue such as stem bases and leaves of cereals.

The washing action of the rain can also bring spores to earth. Both wettable and unwettable spores may be effectively brought out of the air by rain droplets. If rain droplets containing spores land on a plant, the spores are distributed over the plant surface in the trail of water. Rain plays an important part in the dispersal of air-borne conidia and, in addition, many fungi are specifically adapted for dispersal in water droplets. Spores of both *Septoria nodorum* and *S. tritici* are produced in flask-shaped cases, pycnidia. Under conditions of high humidity or free water, the pycnidia exude conidia in a gelatinous matrix,the whole being called a cirrus. This mucilaginous spore mass is ideal for dispersal by raindrops. *Septoria* pycnospores are effectively spread from lesions on lower leaves, vertically up the plant to healthy upper leaves and horizontally from plant to plant during rainfall. Heavy rainstorms appear to be particularly effective in the vertical dispersal of *Septoria* spores and disease epidemics often follow periods of heavy rain in which larger raindrops predominate (Shaw, 1987). In addition, Fitt & Lysandrou (1984) have shown that the numbers of conidia of the cereal eyespot pathogen *Pseudocercosporella herpotrichoides* dispersed in splash droplets increase with increasing raindrop diameter. Transport of water-dispersed pathogens over longer distances may occur by a combination of rain and wind. If spores are taken up in very small water droplets, i.e. less than 20 μm, they may contribute effectively to the air spora and remain in the air for some time.

It is apparent that in several cases, fungal spores are adapted specifically to air or water disperal. However, many species of fungi are able to use any available means (rain, wind or a combination of both) for dispersal through the air.

Survival

Pathogens in both temperate and tropical climates face two main problems related to climate. The first problem is that of extreme climatic conditions: harsh freezing winters in temperate areas and hot dry months in the tropics; the second problem, related indirectly to climate, is that of availability of the host. Pathogens of annual crops need to survive the period when their host is not being cultivated. However, in areas of intensive agriculture, the increasing popularity of, for example, winter-sown cereals, has ensured

that some host material is available for most of the year. Pathogens
of perennial crops have perhaps less of a survival problem although
they may need to adapt to the inactive host out of the main growing
season.

Tarr (1972) categorised five main methods by which fungi survive
or are carried over the harsh season: 'perennial infection, infected
crop residues, alternative host plants, infected planting material
and soil-borne inoculum'. Successful survival of fungi by one or
more of these means will ensure an initial source of inoculum ready
to infect the succeeding crop.

Perennating pathogens

Some pathogens can survive the winter, often in a dormant
condition, on a suitable host plant. It is perhaps easiest for perennial
hosts to be perennially infected. The dormant mycelium of powdery
mildew of apples caused by *Podosphaera leucotricha* may survive
the winter within apple buds. Annual hosts, however, may harbour
pathogens over the winter. Perhaps the most common example of
this is on cereals, notably winter barley. Early-sown autumn crops
of winter barley frequently suffer severe attacks of powdery
mildew. Mildew will generally be killed by temperatures below 0 °C;
however, some mildew may survive as dormant mycelium in winter
barley tissue. Brassica species may act as overwintering hosts for a
number of plant pathogens. Cabbages, brussels sprouts and oilseed
rape, to name but a few, may be found throughout the year in many
areas of the UK. Most pathogens on brassicas are not species-
specific and fungi such as *Peronospora parasitica* (causing downy
mildew) may overwinter on any available brassica crop.

Pathogens in crop residues

Harvesting processes involving annual crops inevitably leave
behind some form of debris, for example cereal stubble and roots. If
these are not either burned or ploughed in soon after harvest, they
may provide material on which pathogens can overwinter. The
extent to which residues are significant for overwintering will
depend on the degree of colonisation, the inherent capability of the
pathogen to survive by competing with saprophytes for the now dead
plant tissue, the ability to produce specialised resting bodies, and
the weather. In relation to the quantity of inoculum on the debris,
diseases such as take-all (caused by *Gaeumannomyces graminis*)
and eyespot (caused by *Pseudocercosporella herpotrichoides*) on
cereals tend to build up in successive cereal crops over years. Each
successive year results in more cereal stubble inoculum ready to

infect the next year's crop. There is a limit to this build-up of
inoculum in the case of take-all, which will be discussed later.

The competitive saprophytic ability of fungi is the summation of
the physiological characteristics which makes for success in the
competitive colonisation of dead organic substrates. The subject
has been covered comprehensively by Garrett (1970). Relatively
specialised pathogens, such as *G. graminis*, which attacks only
cereals, are generally considered not to have high competitive
saprophytic ability. If soil is relatively warm and humid, ideal for
decomposition of dead wheat roots and stubble, much *G. graminis*
inoculum will die out within a year. It cannot survive in the soil as a
free-living saprophyte for any length of time and can be controlled
by crop rotation. Relatively unspecialised pathogens such as some
Fusarium spp., which may attack cereal stem bases, can grow in the
soil, or produce resting bodies, chlamydospores, once their
cereal stem base food source rots away. It is therefore difficult to
control *Fusarium* foot rot of cereals by rotation. The production of
resting bodies by fungi will be considered under soil-borne
inoculum (p. 27).

An example of the effect of weather on overwintering of
pathogens on crop residues has already been mentioned in the case
of take-all. Warm wet weather after harvest will result in the rapid
build-up of saprophytic fungi on debris and this may reduce the
populations of some pathogens. Extreme cold may also directly kill
pathogens on debris.

A very important category of contaminated crop residues is
self-sown or 'volunteer' plants. Again it is inevitable that during any
harvesting operation, some seed is shed or some plants are left
behind in the field. Unless these plants are dealt with, either by
ploughing or by killing with herbicides, they can act as excellent
hosts for fungi over the winter. Volunteer barley plants, for
example, are well established during the autumn months and their
soft green tissue is very prone to diseases like mildew, net blotch and
Rhynchosporium. There may even be some infection by yellow or
brown rust if temperatures do not drop too quickly during autumn
and humidity is high. Even though much of this inoculum will be
killed out by winter frosts, some may survive depending on the
quantity of inoculum on volunteer plants. Some volunteers may
grow from infected seed and therefore carry a high risk of becoming
infected themselves. *Alternaria brassicicola* and *A. brassicae*, the
causes of dark leaf and pod spot of oilseed rape, are seed-borne
fungi. Infection of oilseed rape pods by these fungi, as well as
resulting in dark lesions, causes premature ripening and splitting of

pods. Shed seed may well be infected by *Alternaria* and give rise to infected volunteer plants.

Alternative host plants

Alternative hosts suitable for overwintering may be readily available for pathogens with a wide host range, e.g. pathogens of brassicas. Generally, plants which act as alternative hosts are taxonomically related to the major host plant. Rhizomatous grasses such as *Elymus* (*Agropyron*) *repens* can act as alternative hosts for the take-all pathogen of cereals *Gaeumannomyces graminis*. A slightly different situation occurs in some of the rust fungi, where two hosts are involved (alternate hosts). For example, the alternate hosts of *Puccinia graminis* f.sp. *tritici* are cereals and barberry, which are taxonomically unrelated. *Puccinia graminis* f.sp. *tritici*, which causes black stem rust of cereals, completes its life cycle on the barberry bush. Hence, rather than being used merely as a means of survival, the alternate host is essential for the completion of the life cycle.

Infected or contaminated planting material

Spores and hyphae of many fungi can be borne on seeds or other planting material such as tubers, bulbs or cuttings. The whole area of seed pathology is an important subject covered by Neergaard (1979) in his book *Seed Pathology*. Pathogens may have an intimate relationship with the seed of their host. For example *Ustilago nuda* (causing loose smut of wheat and barley) overwinters in the seed embryo as resting mycelium. The relationship may be less specialised. Some pathogens may contaminate the seed coat, for example *Septoria nodorum*, the cause of glume blotch and leaf spot of wheat. Others produce specialised resting bodies which are not attached to the seed at all but simply contaminate the seed sample. A good example of the latter is the sclerotia produced by *Sclerotinia sclerotiorum*, which causes stem rot of oilseed rape. Some pathogens can overwinter on seed potato tubers, e.g. *Phytophthora infestans*. Should infected potatoes remain undetected and be planted alongside healthy tubers, spread of late blight is likely. *Peronospora destructor*, the causal agent of downy mildew of onion, is an example of a pathogen which can overwinter on bulbs as resting mycelium.

The viability of fungal propagules on seed and planting material generally falls during a few months' storage. There is not necessarily a direct correlation between the percentage of contaminated planting material and the percentage of diseased plants, as some plants may escape infection and others may die. Seed of oilseed rape which is heavily contaminated by *Alternaria brassicae* or *A. brassicicola* may rot prior to or shortly after germination.

Soil-borne inoculum

Pathogens can exist in the soil as free-living saprophytes, as discussed under pathogens in crop residues (p. 24); or pathogens may produce specialised overwintering bodies. Such resting bodies may be produced sexually (zygospores, oospores, some ascospores) or asexually (chlamydospores). Sexual fusion occurs within some resting bodies (teliospores) and some consist of a compacted mass of vegetatively produced hyphae (sclerotia) (Figure 2.3). Some fruiting bodies such as pycnidia, perithecia and cleistothecia embedded in their host tissue may also act as overwintering

Figure 2.3. Specialised fungal resting bodies.

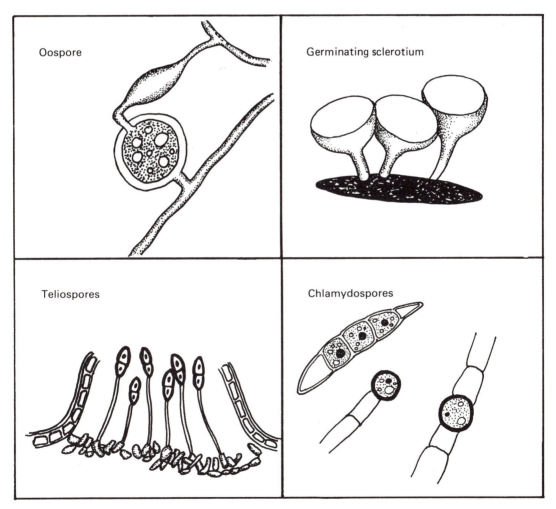

structures. An effective overwintering body must (i) be resistant to extremes and fluctuations of temperature, (ii) be resistant to drying out or waterlogging, (iii) have protection against attacks from other soil organisms such as saprophytic fungi and nematodes and (iv) remain dormant until a suitable host becomes available. An additional useful quality would be germination of the propagule in response to the specific host only. If all resting propagules exhibited all these qualities, the plant pathologist's workload would be increased considerably. Shortcomings in resting propagules may be overcome by mass production. Many propagules germinate in the presence of non-hosts, but in some cases this can be beneficial to the pathogen. Chlamydospores of *Fusarium solani* f.sp. *phaseoli*, the cause of bean root rot, may germinate in the presence of a non-host. However, the cells of the limited mycelium round off to form more chlamydospores, effectively increasing the level of inoculum in the soil.

Some resting propagules have been demonstrated to be very durable in soil. Oospores of *Peronospora destructor* can survive up to 8 years in the soil and resting spores of *Plasmodiophora brassicae* over 20 years. Some sclerotia formed by fungi may survive in the soil for 3 years or more. Fruiting bodies such as cleistothecia and pycnidia have more difficulty surviving for long periods. Undoubtedly pycnidia of *Septoria tritici*, which can be present in large numbers on dead wheat leaves in the autumn, will overwinter successfully and initiate infection of winter wheat in the spring. However, the significance of cleistothecia in overwintering of mildew is more debatable. Cleistothecia of *Erysiphe graminis* (the cause of powdery mildew of cereals) release ascospores primarily in the autumn, when autumn-sown cereals are emerging; fewer survive the winter, to infect cereals in the spring. It is likely that powdery mildews of woody hosts do not have functional cleistothecia and survive winters inside buds as mycelium. For example, Jackson & Wheeler (1974) concluded that very few cleistothecia of *Sphaerotheca mors-uvae* (blackcurrant powdery mildew) overwintered successfully. Soil fauna, freezing temperatures and depletion of food reserves all contribute to the degeneration of the cleistothecia when they do form.

Bacteria as plant pathogens

General characteristics

Bacteria are primitive organisms classified as prokaryotes, i.e. their nuclear material is not separated from the cytoplasm by a nuclear membrane and there is no mitotic apparatus. Most of the genetic information in a bacterial cell is carried on a single chromosome with double-stranded DNA in a closed circular form. In addition, some bacterial cells contain extrachromosomal DNA as plasmids. These are circles of DNA which replicate independently of the chromosomes and are important for several reasons. Genetic material relating to pathogenicity, tumour formation and resistance to chemicals such as antibiotics may be located on the plasmids; plasmids can pass from one bacterial cell to another with ease. Bacterial cells have either no organelles or poorly developed organelles, unlike eukaryotic cells of plants and animals. Almost all bacteria are unicellular and one of three shapes; round (coccus, spherical or ovoid); rods (cylinders) or spirals (helices). All plant pathogenic bacteria are rod-shaped, 0.5–3.5 μm in length and 0.3–1.0 μm in diameter. About 200 out of the total 1600 known bacterial species are recognised as plant pathogens. Most of the remaining species are beneficial to man as saprophytic decomposers although some important bacteria are pathogenic to man and other animals.

The protoplast of a bacterium is surrounded by a cell wall which may take up as much as 25% of the cell volume. This tough cell wall is distinct from the outer cytoplasmic membrane. Outside the cell wall is a slime layer. The nature of this layer depends upon the species of bacteria and the conditions under which they are grown. A thick, well developed slime layer is called a capsule. Many plant pathogenic bacteria have one or more flagella, allowing some limited movement. Flagella may occur at one or both ends (polar) or all over the surface of the bacterium (peritrichous).

All bacteria causing plant diseases are facultative parasites which grow readily in culture but are often slower growing than saprophytic bacteria.

The *Actinomycetes* is now classified as a group of bacteria. This group was previously classified as fungi or bacteria because the organisms produce individual small bacteria-like spores and fungus-like vegetative mycelium.

Reproduction

Bacteria reproduce by a process known as binary fission. This is an asexual process in which a transverse wall develops across the middle of the cell. When new cell wall material has been laid down, the cells separate. During this process the DNA, which is in the form of a circular chromosome, condenses into an amorphous mass which elongates and becomes dumb-bell shaped before it divides into two equal pieces, the 'nuclei' of the daughter cells. Under favourable conditions a bacterium can reproduce every 20 minutes and thus the growth of the population is logarithmic. However, bacteria seldom maintain logarithmic growth for long periods. The growth of bacterial populations is normally limited by exhaustion of available nutrients or the accumulation of toxic metabolic waste products. Nevertheless, it is clear that, given optimum conditions for growth (especially high temperatures), bacteria can reproduce much more quickly than fungi and consequently colonisation and destruction of tissue by bacteria may be rapid.

Taxonomy

The taxonomy of bacteria is a very complex field. The names of bacteria are governed by the International Code of Nomenclature of Bacteria (Lapage *et al.*, 1975; see also Bradbury, 1986). One of the additional complications, as far as plant pathogenic bacteria are concerned, is the existence of pathovars of a number of important species. A pathovar is usually confined to one host and has been defined as 'a subdivision of a species distinguished by common characters of pathogenicity, particularly in relation to host range' (Talboys *et al.*, 1973). Hence the cause of bacterial diseases of many crops, including beans and fruit trees, *Pseudomonas syringae*, has approximately 120 pathovar types. Many guidelines for describing bacteria are used in *Bergey's Manual of Systematic Bacteriology*, the authoritative text on classification of all known bacteria (Kreig & Holt, 1984). Symptomology is the starting point for diagnosis of plant pathogenic bacteria; a comprehensive key based on symptoms and host genus is contained in Lelliot & Stead (1987). In isolation from the host, simple observations such as the colonial appearance on agar and the Gram stain reaction are the usual starting points in identification. Precise identification is, however, a lengthy process and features used include morphology, other staining reactions, oxygen requirements, physiological characters biochemical characters, sensitivity to antibiotics and bacteriocins, phage relationships, serological relationships, disease–host range, protein and fatty acid profiles and nucleic acid studies. A summary of the main characteristics of plant pathogenic bacteria is given in Table 2.1.

Dispersal

There is considerably less information about the dispersal of plant pathogenic bacteria than of fungi, but it is convenient to categorise their modes of dispersal in a similar manner.

Table 2.1. *Main characteristics of plant pathogenic genera of bacteria and typical symptoms produced in their hosts*

Genera	Main characteristics	Typical symptoms
Agrobacterium	The bacteria are rod-shaped, $0.8 \times 1.5–3\,\mu m$. Motile by means of 1–4 peritrichous flagella; when only one flagellum is present it is more often lateral than polar. When growing on carbohydrate-containing media the bacteria produce abundant polysaccharide slime. Colonies are non-pigmented and usually smooth. These bacteria are rhizosphere and soil inhabitants.	Gall diseases of many genera
Clavibacter (*Corynebacterium*)	Straight to slightly curved rods, $0.5–0.9 \times 1.5–4\,\mu m$. Sometimes they have irregularly stained segments or granules and club-shaped swellings. Generally non-motile, but some species are motile by means of one or two polar flagella. Gram-positive.	Potato ring rot, tomato canker and wilt, fruit spot, fasciations
Erwinia	Straight rods, $0.5–1.0 \times 1.0–3.0\,\mu m$. Motile by means of several to many peritrichous flagella. Erwinias are the only plant pathogenic bacteria that are facultative anaerobes. Some Erwinias do not produce pectic enzymes and cause necrotic or wilt diseases ('*amylovora*' group), while others have strong pectolytic activity and cause soft rots in plants ('*carotovora*' group).	Potato blackleg and soft rot, wilt and soft rot of other genera

Table 2.1 (*cont.*)

Genera	Main characteristics	Typical symptoms
Pseudomonas	Straight to curved rods, $0.5–1.0 \times 1.5–4\,\mu m$. Motile by means of one or many polar flagella. Many species are common inhabitants of soil, or of fresh water and marine environments. Most pathogenic *Pseudomonas* spp. infect plants; few infect animals or humans. Some plant-pathogenic *Pseudomonas* species, e.g. *Ps. syringae*, are called fluorescent pseudomonads because, on a medium of low iron content, they produce yellow-green, diffusible, fluorescent pigments. Others, e.g. *Ps. solanacearum*, do not produce fluorescent pigments and make up the non-fluorescent pseudomonads.	Leaf spots, wilts and galls of many genera
Xanthomonas	Straight rods, $0.4–1.0 \times 1.2–3\,\mu m$. Motile by means of a polar flagellum. Growth on agar media usually yellow. Most are slow growing. All species are plant pathogens and are found only in association with plants or plant materials.	Leaf spots, cankers and rots of many genera
Streptomyces	Slender, branched hyphae without cross walls, $0.5–2\,\mu m$ in diameter. At maturity the aerial mycelium forms chains of 3–many spores. On nutrient media, colonies are small (1–10 mm diameter) at first with a rather smooth surface	Common scab of potato, sugar beet, carrot, and other crops

Table 2.1. (*cont*)

Genera	Main characteristics	Typical symptoms
	but later with a weft of aerial mycelium that may appear granular, powdery, or velvety. The many species and strains of the organism produce a wide variety of pigments that colour the mycelium and the substrate; they also produce one or more antibiotics active against bacteria, fungi, algae, viruses, protozoa, or tumour tissues. All species are soil inhabitants. Gram-positive.	
Xylella	Mostly single, straight rods, 0.3 × 1–4 µm, producing long filamentous strands under some cultural conditions. Colonies small, with smooth or finely undulated margins. Gram-negative, non-motile, aflagellate, strictly aerobic, non-pigmented. Nutritionally fastidious, requiring specialised media. Habitat is xylem of plant tissue.	Alfalfa dwarf, ratoon stunting, plum leaf scald

(From Agrios, 1988b.)

Autonomous

Most plant pathogenic bacteria develop primarily in their hosts, but almost all have a saprophytic, soil-borne phase. Hence it is possible to disperse soil-borne bacteria in a similar way to soil-borne fungi, during cultivation. In addition, a typical farming practice such as irrigation of potatoes may disperse *Erwinia carotovora* subsp. *atroseptica*, the cause of blackleg of potato haulms.

Insects and other animals
Insects that feed on plants by sucking or biting frequently transmit
bacteria on their mouthparts. However, in most cases of insect
dispersal of bacteria, there is little evidence of more than a passive
relationship between the bacteria and the insect. It is rare for insects
to be the sole means by which bacteria are dispersed. The cause of
fire blight of apple and pear, *Erwinia amylovora*, is spread mainly
by bees, but flies, aphids, ants, beetles and wasps as well as
rain-splash and contaminated pruning equipment can also spread
the pathogen. There are a few examples of bacteria which are much
more dependent upon insects for their dispersal. In the USA,
E. stewartii, the cause of bacterial, or Stewart's, wilt of maize, is
spread primarily by flea beetles, although it is also seed-borne. The
cucumber wilt organism *E. tracheiphila* is transmitted almost
entirely by *Diabrotica* spp., the cucumber beetles. The bacteria are
apparently unable to overwinter in any location other than
hibernating adult insects. Little is known about long-distance
transport of bacteria in insects. However, there is one particular
example where bacteria which usually spread slowly seem to have
dramatically increased their speed of dispersal by association with
an insect. *Pseudomonas solanacearum*, the cause of diseases of
many tropical crops including 'Moko' disease of bananas, has
spread rapidly in South America on bananas by insect transmission
between inflorescences.

There are a few examples of nematodes being implicated in the
spread of bacterial diseases. *Clavibacter tritici* (*Corynebacterium
tritici*), the cause of 'Tundu' disease of wheat, is unable to produce
symptoms of disease in the absence of the nematode *Anguina tritici*,
which presumably acts as a vector. Any animals which move
through a diseased crop into a healthy one are, as with fungi,
potential vectors for bacterial plant pathogens.

Transmission of bacteria over long distances is usually effected by
man. Any contaminated planting material such as seed and cuttings,
together with soil and other material containing bacteria, provides
potential means of spread.

Air and water
Bacterial dispersal in air is usually in water droplets (aerosols).
Bacteria may be dispersed by air alone on contaminated debris or
soil particles and, if such material is small enough, bacteria may be
carried over some distance. Rainwater can wash, disperse and
distribute bacteria on one plant, and splash bacteria from plant to
plant. Rain also washes bacteria onto lower parts of the plant and

into the soil. Disease epidemics are often associated with damaging storms. Bacterial aerosols may be taken up into the air and dispersed over longer distances by air currents.

Survival

Perennating pathogens
The bacterium *Erwinia amylovora*, the cause of fire blight of apple and pear, overwinters in its perennial hosts within the wood of the trees at the margins of old cankers.

Pathogens in crop residues
Trash and debris, contaminated with bacteria on or shallowly buried in the soil, can act as good overwintering sites for bacteria. Plant pathogenic bacteria are generally not highly competitive saprophytes and most, possibly with the exception of *Agrobacterium*, decline in numbers in the soil. If infected crop residues are deeply buried by ploughing, plant pathogenic bacteria probably do not survive. However, there exist examples of bacteria surviving on trash, including *Pseudomonas syringae* pv. *tabaci* (wildfire of tobacco), *Xanthomonas campestris* pv. *phaseoli* and *Pseudomonas syringae* pv. *phaseolicola* (bacterial blights of beans), and *Pseudomonas solanacearum* (bacterial wilt of solanaceous plants).

Alternative host plants
It is difficult to find reports of bacteria surviving on alternative host plants but this must happen in some cases. For example, it is possible that *Xanthomonas campestris*, the cause of black rot of crucifers, could overwinter on any cruciferous species available.

Infected or contaminated planting material
Bacteria commonly survive the winter on seed. The bacterial mucilage can dry on the seed coat and some bacteria can survive for several years in this state. The bacterial blights of beans caused by *Xanthomonas campestris* pv. *phaseoli* and *Pseudomonas syringae* pv. *phaseolicola* survive on seed as well as on debris. *Erwinia carotovora* subsp. *atroseptica*, the cause of blackleg of potatoes, overwinters on seed potatoes. In this particular case the bacteria survive undetected in a latent form deep in the potato lenticels. Any contaminated material taken into store for planting in the subsequent year is a potential overwintering source of inoculum.

Soil-borne inoculum

Few plant pathogenic bacteria live freely in the soil as saprophytes. Although *Agrobacterium tumefaciens*, the cause of crown gall disease of a range of hosts, was generally considered to be a true soil inhabitant, recent evidence suggests that pathogenic strains survive either in tumours on living plants or in close proximity to plant material. Plant pathogenic bacteria have no complex specialised survival structures comparable to those of fungi and soil-borne inoculum generally takes the form of infected crop residues.

Mycoplasma-like organisms as plant pathogens

General characteristics

Mycoplasma-like organisms (MLOs) are similar to bacteria in that they are prokaryotes and lack an organised bounded nucleus. However, MLOs differ from bacteria in that they lack rigid cell walls, being surrounded only by unit membranes. This allows them to be highly pleomorphic and they assume a vast array of shapes and sizes (Figure 2.4). The diameter of MLOs varies from 300 nm to 1 μm and MLOs include the smallest known cells able to multiply independently of other living cells. As recently as 1967, workers first

Figure 2.4. Electron micrograph of the MLO clover phyllody in clover (×70 000). (P. G. Markham, John Innes Institute.)

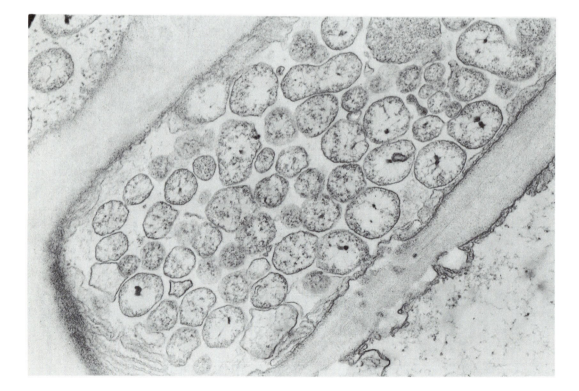

discovered MLOs in the phloem of plants which were suffering from yellows-type diseases. Since then, 70 diseases affecting over 300 genera of plants have been attributed to MLOs. It is difficult, if not impossible, to grow MLOs on artificial growth media and, remarkably, impossible to reproduce disease symptoms in healthy plants by artificial inoculation.

At present few if any diseases of economic significance to British agriculture are known to be caused by MLOs. However, it is possible that as more detailed examination of some virus diseases is undertaken, MLOs may be implicated as part of the problem.

Reproduction
MLOs reproduce by budding and binary fission.

Taxonomy
The MLOs are included in the class Mollicutes. The class has one order, the Mycoplasmatales, which is divided into three families each containing one genus, *Mycoplasma*, *Acholeplasma* and *Spiroplasma*, respectively. These genera are separated on their sterol requirements for growth. Diseases caused by MLOs are named after the most striking symptom produced in the host. As well as the yellows-type symptom, MLOs can produce growth abnormalities such as green flowers and proliferation of axillary buds; affected plants also often display a loss of vigour. Sometimes infection by MLOs may cause flowers to develop into leafy structures (phyllody). MLO diseases include aster yellows, clover phyllody, citrus stubborn and potato witches broom.

Dispersal (transmission)
MLOs are found almost exclusively in the phloem of infected plants and require a vector for their dispersal. As with plant virus diseases, insects are primary vectors of MLOs. The majority of MLOs are transmitted by leafhoppers although psyllids, treehoppers, plant-hoppers, aphids and mites have also been implicated. The vector acquires the MLO while feeding on contaminated phloem. An incubation period follows during which time the vector cannot transmit the MLO. The incubation period lasts between 10 and 45 days depending on the temperature. During this time the MLO multiplies in the vector, particularly in the salivary glands and the haemolymph. The vector now remains infective for the remainder of its life, but MLOs are not passed on through the eggs of a contaminated insect. In addition to insect dispersal, MLOs may be

transmitted by grafting and by the parasitic plant, dodder (*Cuscuta* spp.).

Survival

Many MLOs and their vectors have a wide host plant range. For example, aster yellows occurs in carrot, lettuce, onion, spinach, potato, barley, flax, aster, gladiolus, tomato, celery and phlox and is transmitted by a number of different leafhoppers. In addition to cultivated plants, the MLO causing aster yellows has been found in a number of weeds including thistle, wild carrot, dandelion and field daisy. It is obvious that there is no shortage of plant host material for this MLO at any time of year. A number of MLOs attack perennial plants. Pear, peach and citrus trees are liable to MLO-induced diseases. One of the most significant diseases of citrus trees is citrus stubborn disease caused by *Spiroplasma citri*. In some Mediterranean countries the disease has resulted in so many unproductive trees that citrus production is uneconomic. It is probable that in perennial plants the organism survives in a dormant phase during the host's dormant period.

Viruses as plant pathogens

General characteristics

A virus has been defined in two ways by Hollings (1983*a*): 'An obligate parasite of sub-microscopic size, with one dimension smaller than 200 nm' and, alternatively, 'A set of instructions to a suitable host organism to synthesise more virus'.

These definitions highlight the two most important features of a virus: it is very small; and it needs a living cell in order to reproduce. It is this latter characteristic of viruses that makes them particularly significant: because they require a host for replication, all viruses are parasites. Viruses cause diseases in man and other animals, plants, fungi, bacteria, and even MLOs. Of the 1000 or more well characterised and classified viruses, over 500 cause diseases in plants and a plant may be simultaneously affected by more than one virus.

As well as being amongst the smallest of the plant pathogens, viruses are the most simply constructed. They do not consist of cells, but of particles made up of a nucleic acid core of either single-stranded or double-stranded RNA or DNA, surrounded by a protein coat. Using the transmission electron microscope, virus particles can be seen to be elongated (rigid rods or flexuous threads), shorter bacillus-like rods (rhabdoviruses), or spherical (isometric or polyhedral) (Figure 2.5). The protein coat consists of normally identical sub-units which form the external coating of spherical particles or the spirally arranged sub-units of elongated particles. A few plant viruses have been found to consist of more

Figure 2.5. Different shapes of virus particles. (i) Flexuous threads of Turnip mosaic virus (×79 000), (ii) short bacillus-like rods of Broccoli necrotic yellows virus (×191 000), (iii) isometric particles of Cucumber mosaic virus (×160 000). (C. M. Clay, Institute of Horticultural Research, Wellesbourne.)

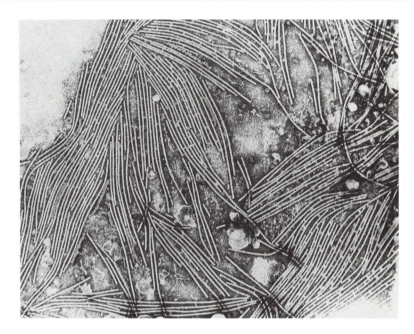

Fig. 2.5. (ii)

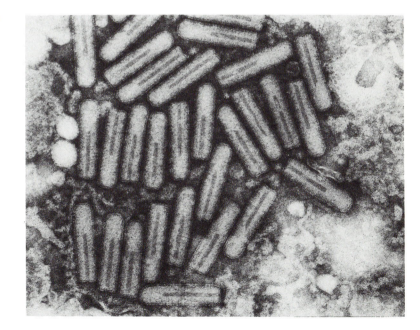

Fig. 2.5. (iii)

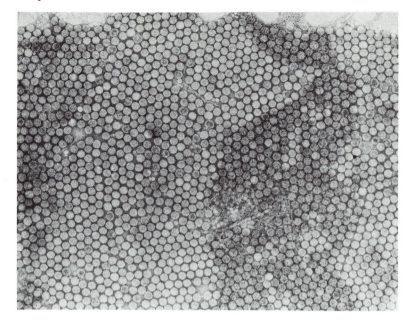

than one component or particle, for example, alfalfa mosaic virus consists of five components, all of different sizes.

Reproduction (replication)

The reproduction of viruses is a complex biochemical process which will not be dealt with in depth. Viral replication is unique in that the virus uses the host cell's synthetic machinery for its own replication. The first stage is release of the viral nucleic acid (most commonly single-stranded RNA) from the protein coat. The virus then induces the formation of RNA polymerase enzymes in the host cell. The information for the synthesis of these enzymes may come partly from the host and partly from the virus. The polymerase catalyses the formation of RNA which is a mirror image of the RNA in the virus particle. The double-stranded RNA thus formed now separates to produce more of the original viral RNA and its mirror image which acts as a template for further virus RNA synthesis. When a virus consists of more than one particle, replication is more complex, but all particles must be present in one host cell for replication to occur successfully.

The new viral nucleic acid in the host cell induces the formation of further viral protein sub-units by host ribosomes 'reading' the appropriate message encoded in part of the virus RNA, which carries instructions for the production of other proteins in the host cell. Interference with such a fundamental part of the host cell's metabolism results in some of the symptoms of virus diseases which include mosaics, mottling, growth abnormalities and distortions, chlorosis, and stunting.

Taxonomy

Several attempts have been made since the discovery of viruses to devise a rational, taxonomically sound classification system. Most viruses are still commonly known according to the host from which they were first isolated and the most striking symptoms they produce. According to Walkey (1985), 27 groups of plant pathogenic viruses have been so far accepted by the International Committee on Taxonomy of Viruses (Table 2.2).

Table 2.2. *Groups of plant pathogenic viruses*

Group name and derivation	Examples
1. Luteovirus (Luteus = yellow)	Barley yellow dwarf, Beet western yellows, Pea leaf roll, Turnip yellows
2. Maize chlorotic dwarf	Rice tungro
3. Sobemovirus (*Southern bean mosaic virus*)	Southern bean mosaic, Cocksfoot mosaic, Turnip rosette
4. Tobacco necrosis	Tobacco necrosis, Cucumber necrosis
5. Tombusvirus (*Tomato bushy stunt*)	Tomato bushy stunt, Carnation Italian ringspot
6. Tymovirus (*Turnip yellow mosaic*)	Turnip yellow mosaic, Andean potato latent
7. Comovirus (*Cowpea mosaic*)	Cowpea mosaic, Bean pod mottle, Andean potato mottle
8. Dianthovirus (Carnation ringspot virus group)	Carnation ringspot, Red clover necrotic mosaic
9. Nepovirus (*Nematode-borne polyhedral particles*)	Tobacco ringspot, Raspberry ringspot, Tomato ringspot
10. Pea enation mosaic	Pea enation mosaic only
11. Alfalfa mosaic	Alfalfa mosaic only
12. Bromovirus (*Brome mosaic*)	Broadbean mottle, Cowpea chlorotic mottle

(*continued overleaf*)

Table 2.2. (*cont.*)

Group name and derivation	Examples
13. Cucumovirus (*Cucu*mber *mo*saic)	Cucumber mosaic, Peanut stunt
14. Ilarvirus (*I*sometric *la*bile *r*ingspot)	Tobacco streak, Apple mosaic
15. Velvet tobacco mottle	Velvet tobacco mottle, Lucerne transient streak
16. Tobravirus (*Tob*acco *r*attle)	Tobacco rattle, Pea early browning
17. Tobamovirus (*Toba*cco *mo*saic)	Tobacco mosaic, Tomato mosaic, Potato mop-top, Beet necrotic yellow vein
18. Hordeivirus (Hordeum = barley)	Barley stripe mosaic
19. Potexvirus (*Pot*ato virus *X*)	Potato X, Cassava common mosaic, Clover yellow mosaic
20. Carlavirus (*Car*nation *la*tent)	Carnation latent, Pea streak, Cowpea mild mottle, Alfalfa latent
21. Potyvirus (*Pot*ato virus *Y*)	Potato Y, Bean common mosaic Beet mosaic
22. Closterovirus (Kloster = spindle)	Beet yellow stunt, Citrus tristeza
23. Rhabdovirus (Rhabdos = rod)	Lettuce necrotic yellows, Barley yellow striate mosaic, Beet leaf curl
24. Tomato spotted wilt	Tomato spotted wilt only
25. Reovirus (a) Phytoreovirus (*R*espiratory *e*nteric *o*rphan) (b) Fijivirus	Wound tumour, Rice dwarf Fiji disease, Maize rough dwarf
26. Geminivirus (Gemini = twins = paired virus particles)	Maize streak, Beet curly top
27. Caulimovirus (*Cauli*flower *mo*saic)	Cauliflower mosaic, Dahlia mosaic

(After Walkey, 1985.)

Although there is no definitive list of biological or biochemical characters which can be used to classify viruses, Hollings (1983*b*) lists 10 areas where as much information as possible should be gathered in order to describe a virus fully.

1. Experimental transmission: can the virus be mechanically transmitted or does the plant parasite dodder have to be used?
2. The host range and symptoms, both natural hosts in the field and a range of experimental test plants, including the effect of temperature, light and virus concentration on symptoms.
3. The virus's properties in the sap, incuding the dilution end point (dilution of sap at which symptoms fail to appear), thermal inactivation point, survival at room temperature and duration of survival at freezing point.
4. The possibility of transmission of the virus through seed and pollen.
5. The transmission by vectors, including detailed information about latent periods, persistence and transmission (see Dispersal).
6. The method of purification, including suitable buffers, precipitation agents and centrifugation regimes.
7. The properties of purified preparations, including size and shape of virus particles, sedimentation properties, density, chemical composition and nucleic acid content.
8. Tests for relationships with other viruses: serological (comparison of reactions of viral antisera), plant protection (protection of infected plants against experimental infection with other viruses).
9. Detailed cytological studies of infected tissue for inclusion bodies.
10. The ability to eliminate a virus by heat treatment and/or meristem tip culture.

Clearly, describing and attempting to classify a virus is a specialised field requiring much complex laboratory equipment and expertise.

Dispersal (transmission)

Understanding how viruses are dispersed or transmitted is essential for proof of pathogenicity. Plant viruses rarely emerge from plants spontaneously and although they may be passively dispersed by wind, water or air from infected tissue, they are usually dispersed by vectors and can enter plants only via wounds. The mode of transmission has been identified for most virus diseases although

there are still several, particularly virus diseases of fungi, whose
mode of transmission remains uncertain.

Autonomous

Autonomous dispersal of plant pathogenic viruses is rare. However, if a virus is present in high concentrations in the sap, the hosts
are grown densely and tissue is easily wounded, natural mechanical
transmission may occur. This is the case for potato virus X where
potato leaves brushing against each other or machinery moving
through a potato crop can damage plants and infected sap may be
transferred from diseased to healthy plants. Transmission of viruses
by natural root grafts may also occur, e.g. apple mosaic virus.

Insects, other animals and fungi (vectors)

This means of transmission has received much attention. Insects in
particular are the most common and economically important
carriers or vectors of viruses and will be discussed first. There are
often complex relationships between insect vectors and viruses and
some of the technical terms which help to describe aspects of these
relationships will be defined.

The **latent period** in a vector/virus relationship is the time from
the start of the vectors' feeding period until the vector can transmit
the virus to healthy plants. The **persistence** of a virus is the time for
which a vector remains infective after leaving the virus source.
Persistence has been divided into: **non-persistent**, when the virus
persists for a few (usually less than 4) hours at approximately 20 °C;
semi-persistent, when the virus persists for between 10 and 100
hours; and **persistent**, when the vector remains infective for more
than 100 hours and in some cases the life of the vector. Other terms
used include **stylet-borne viruses**, which are probably carried on the
stylets (probing mouthparts) of vectors; **circulative viruses**, which
pass through the gut wall into the haemolymph of the vector and
eventually reach the mouthparts of the vector via the saliva but do
not multiply in the vector; and **propagative viruses**, which do
multiply in their vector. These categories are discussed further by
Gibbs & Harrison (1976).

Insect/arachnid transmission The range of arthropods which have
been shown to transmit viruses includes aphids, leaf, plant and
treehoppers, white flies, beetles, thrips, mealy bugs and grasshop-
pers, as well as mites. The most important of these are the aphids;
the great majority of all stylet-borne viruses are transmitted by
aphids. The important virus disease of potatoes resulting in the

symptoms of severe mosaic, rugose mosaic and leaf drop streak is caused by potato virus Y (PVY) and is spread by aphids, primarily the peach potato aphid, *Myzus persicae*. Aphid-borne viruses such as PVY are acquired very quickly by aphids probing diseased tissue and in most cases the virus is transmitted to a healthy plant in a few seconds. However, the persistence of such stylet-borne viruses is short, transmission only lasting a few hours (non-persistent). There are only a few cases of aphid vectors transmitting circulative viruses where latent periods are longer, but in these cases the viruses are more persistent. Potato leaf roll virus, probably now the most important virus disease of potatoes, is of the persistent type. The aphid vector, again mainly *M. persicae*, needs to feed for about 2 hours on an infected plant in order to acquire the virus. A further 24 hours is required before the virus can be transmitted but the aphid then remains infective for a number of days.

The next most important group of virus vectors are the leaf, plant and treehoppers, the most significant of these being the leafhoppers. Many of the yellowing and leaf-rolling symptoms are caused by viruses spread by leafhoppers. All leafhopper-transmitted viruses are circulative and several have been observed to multiply in their vectors (propagative viruses). Some viruses can persist through the moult of the leafhopper and may be transmitted through the egg (transovarial transmission). In addition, most leafhopper transmitted viruses are persistent, often resulting in the leafhopper retaining infectivity for the rest of its life. Normally there is a latent incubation period of 1–2 weeks. Curly top virus causing curly top of sugar beet is spread by the leafhopper *Circulifer tenellus*.

Mites have been recorded as vectors for a relatively small number of viruses. The relationship between a mite vector and a virus is often quite specific. The mite may be the only known vector of a particular virus. Some of the viruses are stylet-borne and others are circulative. The mites *Eriophyes tulipae* and *E. tosichella* transmit wheat streak mosaic virus.

Nematode transmission Two groups of viruses are transmitted by soil-inhabiting plant parasitic nematodes; Nepoviruses and Tobraviruses. There are about 30 plant viruses, transmitted by four genera of nematodes; *Longidorus*, *Xiphinema*, *Trichodorus* and *Paratrichodorus*. In principle, the method of transmission by nematodes is quite simple. The long probing mouthpart (stylet) of root-feeding nematodes picks up the virus from infected plants. Healthy plants are infected by the probing of a contaminated stylet.

There is no evidence for viruses multiplying in nematodes or being transmitted via nematode eggs. An example of a nematode-transmitted virus disease is 'spraing' of potatoes caused by tobacco rattle virus. The vectors of this virus in the UK are *Paratrichodorus pachydermus* and *Trichodorus primitivus*. The virus causes stem mottle in potato haulms and internal brown corky arcs in infected tubers (spraing is a Scottish dialect word for streaks or stripes).

Fungal transmission Viruses transmitted by fungi are of increasing importance in agriculture in the UK. Potentially the most important disease of sugar beet, rhizomania, is caused by beet necrotic yellow vein virus, transmitted by a fungus, *Polymyxa betae*. All fungi which transmit viruses to plants are soil-borne and belong to the lower genera *Olpidium*, *Spongospora*, *Polymyxa* and *Pythium*. These all produce motile zoospores and do not usually cause serious disease attacks. Most of the genera are much more significant as virus vectors than pathogens in their own right. The viruses seem to have the ability to be carried either internally or on the surface of zoospores which infect plants, and they can also survive in the soil within the fungus resting spore, for as long as the spore remains viable. As well as rhizomania of sugar beet, discovered in the UK for the first time in 1987, another increasingly serious virus disease in the UK is barley yellow mosaic virus (BaYMV) which is spread by the soil-borne fungus *Polymyxa graminis*.

Transmission by man Man may be responsible for transmission of viruses during horticultural propagation of plant material (grafting, cuttings, planting bulbs and corms) or by planting virus-infected seed. Most fruit trees and ornamentals are produced by grafting and seed or tubers of many field crops could potentially be produced from infected stock. Man's involvement in virus transmission over short and long distances could, therefore, be highly significant.

Survival
Viruses, like other obligate pathogens or biotrophs, have the problem of finding suitable hosts on which to survive between crops. In addition, a suitable vector must be available for transmission.

Perennating pathogens
Perennial cultivated plants such as fruit trees may act as overwintering sources of viruses. Symptoms of virus infection may not be obvious all year round in perennially infected plants. Several species of perennial weeds act as reservoirs of viruses in the USA.

Pathogens in crop residues

While stubble and debris play an important part in overwintering of fungi, it is very rare for them to act as overwintering sources of virus if plant material is dead. However, living contaminated crop residues such as cereal volunteers and potato ground keepers are important overwintering virus sources. Barley yellow dwarf virus, for example, can overwinter in volunteer wheat, barley and oats.

Alternative host plants

A very common method of overwintering of virus diseases is in alternative hosts. Many viruses have a wide host range which may or may not be related to the primary cultivated host. Barley yellow dwarf virus can overwinter in numerous wild perennial grasses. Weeds can be considered as alternative host plants, for example beet yellows virus overwinters on susceptible weeds such as fat hen and shepherd's purse. An alternative host, as well as being susceptible to the virus, must also be a suitable host for the virus vector.

Infected or contaminated planting material

As far as agriculture is concerned, virus-infected seeds are probably the most significant problem. Of the 100 or so viruses which have been reported as seed-transmissible, the percentage of virus-infected seed from virus-infected plants is generally low. Much virus infection of seeds occurs in the embryo, although seed coat and endosperm have occasionally been found to contain virus. Of the viruses which persist on the seed coat, tobacco mosaic virus has been recorded as surviving for 3 years externally on seeds of tomato. It is perhaps surprising that more viruses are not seed transmitted, particularly as several viruses are transmitted by infected pollen. Infected planting material taken into store for planting the next season, e.g. potato tubers, can potentially be an important overwintering source of virus. Symptoms may not be evident in either mother plants or tubers during the season of infection. However, plants grown from infected tubers may subsequently develop severe and damaging symptoms.

Soil-borne inoculum

Infection of plants by free virus particles in the soil is difficult if not impossible in the field. Viruses must be in a living, active organism such as a nematode or in a living but dormant structure such as a fungal resting spore. As previously discussed, soil-borne fungal resting spores such as those of *Polymyxa graminis*, which harbour barley yellow mosaic virus, may retain their viability and the viability of the virus for many years in soil. As such, virus-contaminated resting spores pose a particularly insidious threat to crop production in some areas.

Further reading

Fungi

Bruehl, G. W. (1987). *Soilborne Plant Pathogens*. New York: Macmillan.

Garrett, S. D. (1970). *Pathogenic Root-Infecting Fungi*. Cambridge: Cambridge University Press.

Hawksworth, D. L., Sutton, B. C. & Ainsworth, G. C. (1983). *Dictionary of the Fungi*. Slough: Commonwealth Agricultural Bureaux.

Ingold, C. T. (1978). Dispersal of micro-organisms. In *Plant Disease Epidemiology*, ed. P. R. Scott & A. Bainbridge, pp. 11–21. Oxford: Blackwell.

Neergaard, P. (1979). *Seed Pathology*, Vols 1 and 2. London: Macmillan.

Tarr, S. A. J. (1972). *Principles of Plant Pathology*. London: Macmillan.

Webster, J. (1980). *Introduction to Fungi*. Cambridge: Cambridge University Press.

Bacteria

Billing, E. (1987). *Bacteria as Plant Pathogens, Aspects of Microbiology 14*. Wokingham: Van Nostrand Reinhold (UK).

Johnston, A. & Booth, C. (ed.) (1983). *Plant Pathologist's Pocketbook*, pp. 30–45. Slough: Commonwealth Agricultural Bureaux.

Lelliot, R. A. & Stead, D. E. (1987). *Methods for the Diagnosis of Bacterial Diseases of Plants, Methods in Plant Pathology* Vol. 2, ed. T. F. Preece. Oxford: Blackwell.

Rhodes-Roberts, M. E. & Skinner, F. A. (ed.) (1982). *Bacteria and Plants. The Society for Applied Bacteriology Symposium Series No. 10*. London: Academic Press.

Mycoplasma-like organisms

(See also references below for virus diseases.)

Markham, P. G. & Townsend, R. (1983). Mycoplasma-like organisms as plant pathogens. In *Plant Pathologist's Pocketbook*, ed. A. Johnston and C. Booth, pp. 79–89. Slough: Commonwealth Agricultural Bureaux.

Viruses

Bos, L. (1983). *Introduction to Plant Virology*. Harlow: Longman.

Gibbs, A. & Harrison, B. D. (1976). *Plant Virology: the Principles*. London: Arnold.

Hill, S. A. (1984). *Methods in Plant Virology*. Oxford: Blackwell.

Hollings, M. (1983). Virus diseases. In *Plant Pathologist's Pocketbook*.

ed. A. Johnston & C. Booth. pp. 46–77. Slough: Commonwealth
Agricultural Bureaux.
Walkey, D. G. A. (1985). *Applied Plant Virology*. London:
Heinemann.

3 How does disease build up?

The build-up, or development of disease will be considered in two sections: firstly, disease development in individual plants (infection and colonisation); and secondly, disease development in populations of plants (epidemiology).

Infection and colonisation of plants

The varied survival and dispersal mechanisms of fungi, bacteria, mycoplasma-like organisms and viruses may be re-stated very briefly: pathogens are often dormant, perhaps on volunteer plants, debris or seed, or floating in air-currents. From this base pathogens have to recommence their life cycle. A pathogen must first encounter and penetrate a susceptible host plant. It must then feed on the host in order to grow and reproduce. Finally, an alternative host must be found or specialised survival structure formed in preparation for winter absence or dormancy of its host. Assuming that the pathogen has been successfully dispersed, germinated and come into contact with the host, the development of the pathogen on the surface of the host, the penetration of the host and the earliest stages of development may all be considered as part of **infection**. As soon as the pathogen starts to depend on its host for nutrients and grow away from the initial site of infection into the surrounding tissue, **colonisation** has occurred.

Infection

The factors which first influence infection of plants are those which affect the germination of pathogen spores. These include temperature, water, light and pH, some of which have already been discussed in the context of the general effect of the environment on disease development (Chapter 1). Exogenous nutrients or inhibitors produced by plants (exudates) and other micro-organisms are also important at this stage.

Effects of plant exudates on infection

The amount and constitution of plant exudates depends on the plant species, variety, the age and part of plant. In roots, substances exude from living cells, dead material and breakdown products. The most important exudates are mineral salts, sugars and amino acids. Exudates from leaves are composed of similar materials. Seeds are known to exude substances freely, again including sugars and amino acids. Exudates may have a tropic or tactic effect on pathogens. Fungi which produce motile zoospores, such as some soil-inhabiting *Phytophthora* spp., have been reported to exhibit chemotaxis towards roots of both host and non-host species and this reaction is often stronger in damaged roots.

Root exudates have their effect in the **rhizosphere** (the zone in soil affected by roots). In the rhizosphere, the germination of dormant fungal resting propagules may be stimulated, reduced or totally inhibited by the action of root exudates. There is evidence that sclerotia of *Sclerotium cepivorum*, the cause of white rot of onions, are stimulated to germinate in the rhizosphere of susceptible *Allium* spp. Resting spores of *Plasmodiophora brassicae* have been shown to break dormancy as a result of an isothiocyanate component of some root exudates.

In shoots, exudates from the cuticle, particularly of leaves, can affect the germination and growth of plant pathogens. Some spores which land on leaves of non-hosts do not germinate. It is thought that plant metabolites may diffuse through the cuticle and inhibit germination. Other spores germinate but fail to colonise plants. This may be a result of the production of specific anti-fungal agents (**phytoalexins**), which is discussed in Chapter 5. There is evidence that the sugars produced in shoot exudates may encourage infection in some circumstances. The formation of swollen hyphal tips, which adhere closely to the plant surface (**appressoria**) and are a pre-requisite for infection by the stem rot pathogen *Sclerotinia sclerotiorum*, appears to be stimulated by shoot exudates in some circumstances.

Effects of other micro-organisms on infection

In the **infection court** (initial site of contact between the pathogen and the surface of the host), the pathogen is rarely alone. Other micro-organisms, which may be pathogenic or purely saprophytic, are likely to be present and can influence the germination and infection by the pathogen. Exploitation of the ability of saprophytes to compete with the pathogen and thereby reduce disease is the basis of **biological control** (Chapter 5). Such competitive saprophytes, which decrease the growth of another micro-organism, are known as **antagonists**. Antagonism has been studied mainly in connection with the effects on soil-borne pathogens. However, many studies of soil-borne Actinomycetes and aggressive saprophytes such as *Trichoderma viride* have not given sufficient evidence to prove that specific changes in the rhizosphere microflora can affect the capability of the pathogen spore to germinate and infect plant roots. Interactions between pathogens and saprophytes on leaf surfaces have been less well studied. Leaves carry micro-organisms such as yeasts and saprophytic fungi of the genera *Cladosporium*, *Aspergillus* and *Penicillium*. Conidia produced by such saprophytes on the surface of leaves may themselves exude substances which may stimulate the growth of bacteria antagonistic to the pathogen. However, such links between saprophytes and pathogens are difficult to substantiate.

There is considerable evidence that interactions between different pathogenic species or different races of the same species can affect infection of plants. Pre-inoculation with an avirulent race of a pathogen such as brown rust (caused by *Puccinia recondita*) or mildew (caused by *Erysiphe graminis*) decreases infection by the virulent pathogen. Finally, interactions between propagules of the same pathogen may influence infection. Self-inhibition and self-stimulation are well recognised phenomena in a number of pathogens. Conidia of *Peronospora tabacina*, the cause of blue mould of tobacco, will not germinate if too densely packed. In contrast, uredospores of some rust fungi may be stimulated to germinate by the production of phenolic substances by the spores themselves.

Penetration of the host
Once a spore has germinated, it must gain access to the plant. It can achieve this in three ways: by direct penetration of the outer layers of plant tissue, by entry through natural openings, or by entering via wounds.

Direct penetration
The outer surface of plants consists of a protective layer of cuticle which is itself covered with wax outgrowths. This barrier must be breached if infection is to occur. The cuticle is composed of a spongy

matrix of cutin with wax, which is continuous over the plant surface except at natural openings and offers a good deal of resistance to any potential invader. Indeed, bacterial and viral plant pathogens cannot penetrate this layer. However, the direct penetration of aerial and underground parts of the plant by fungal pathogens does occur and is effected in similar ways although the layer of cuticle is generally much thinner on roots than stems and leaves. Both biotrophs such as mildews and necrotrophs such as *Botrytis* spp. have the ability to penetrate the cuticle directly. The penetration may be a mechanical or a chemical process or a combination of both. Penetration often involves the formation of an appressorium (Figure 3.1) which brings the fungal hypha into close contact with the plant surface. In addition, an **infection peg**, a structure formed by the deposition of substances such as lignin around a thin hypha, may develop underneath the closely adhered appressorium (Figure 3.1). Penetration is presumably facilitated by a thinner hypha which will penetrate the surface more easily than an unmodified thicker

Figure 3.1. Infection of barley by a conidium of *Erysiphe graminis* f.sp. *hordei*. (S. Archer, Imperial College.)

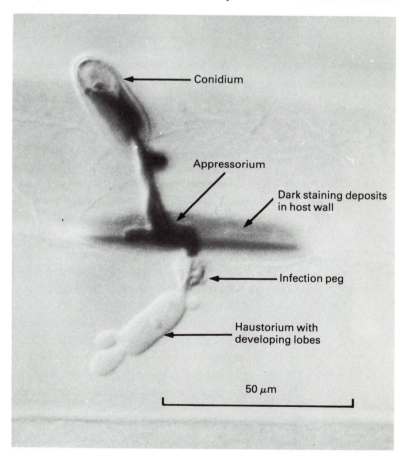

Conidium

Appressorium

Dark staining deposits in host wall

Infection peg

Haustorium with developing lobes

50 μm

hypha. The evidence for purely mechanical penetration includes observation of infection pegs being formed on and penetrating inert substances such as thin gold foil. However, in some diseases there is evidence for degradation of cutin involving an enzyme, cutinase. The role of cutinase in the penetration of pea cuticle by the foot and root rot pathogen *Fusarium solani* f.sp. *pisi* has been confirmed using antibodies against the enzyme. Extracellular cutinase activity has also been demonstrated in other pathogens including *Botrytis squamosa* (leaf rot of onions) and *Colletotrichum graminicola* (anthracnose of maize) (Kolattukudy, 1985). It appears that both physical and enzymatic processes are involved in some cases of cuticle penetration.

Other interesting specialised penetration structures are found in soil-borne pathogens. *Rhizoctonia solani*, the cause of many diseases in crop plants including stem canker and black scurf of potatoes, forms unusual lobed appressoria and infection cushions each composed of a mass of hyphae. It is almost as if a concentration of fungi is required in order to effect penetration of the host by this pathogen. The unique process by which another soil-borne plant pathogen *Plasmodiophora brassicae* enters plants has been described by Williams, Aist & Bhattacharya, (1973). Zoospores of *P. brassicae* become attached to root hairs of cabbage and encyst. A penetration apparatus then forms within the cyst. At the start of penetration, a sock-like structure (**rohr**) emerges from the cyst and closely adheres to the outer wall of the host. A bullet-shaped structure (**stachel**) passes down the rohr and through the root hair wall. Immediately, a stream of cytoplasm is transferred from the cyst to the host. This is a particularly dramatic example of mechanical penetration, where the actual infection process takes only about one second.

Entry through natural openings

The most important natural openings, as far as pathogens are concerned, are stomata. Several downy mildew fungi, including *Plasmopora viticola* (downy mildew of vines) and a number of rust fungi including *Puccinia graminis* (black stem rust of wheat), enter plants almost exclusively via the stomata. Upon germination of spores of such fungi, germ tubes grow towards stomata where, in the rust fungus, a wedge-shaped structure is produced in the stomatal aperture. From this structure a sub-stomatal vesicle is formed from which hyphae grow and further colonise plant tissue. In contrast, during wet weather zoospores of downy mildew fungi swim into stomata where they encyst and then produce hyphae. Stomata also provide one of the main means of entry for bacterial plant pathogens. Entry is probably by a contaminated film of water which is continuous between the leaf surface and sub-stomatal cavity. As such, bacterial entry into plants may be a simple passive process.

The other main natural openings are the lenticels. These loosely packed groups of cells, which occur in bark on woody stems, on secondary thickened roots and in the periderm of potato tubers, provide an ideal opportunity for bacteria and other pathogens to enter plant tissue. *Phytophthora infestans* (causing late blight of potatoes), *Streptomyces scabies* (causing common scab of potatoes) and, perhaps most importantly, *Erwinia carotovora* subsp. *atroseptica* (causing blackleg of potatoes) can enter potato tubers via lenticels. *Erwinia amylovora*, the cause of apple and pear canker, commonly enters leaves via secretory glands (hydathodes) during disease epidemics.

Entry through wounds

If the outer protective layer of intact periderm which covers most of a mature plant is damaged, the plant is vulnerable to a wide range of pathogens. Many wound pathogens are unspecialised facultative parasites which cannot infect undamaged, healthy plants. Wounding occurs in a variety of ways. Wind, rain, frost and pests (including man and his machinery!) can all damage plants. Indeed these factors, together with some of the natural growth processes of plants such as the formation of lateral rootlets and leaf abscission which result in self-inflicted wounds, lead one to believe that there is no such thing as a totally undamaged plant. Clearly there is much potential for pathogens which can take advantage of damaged tissue. Many of these pathogens are most troublesome in stored crops. Fleshy potato tubers and fruit are particularly prone to wound pathogens. Many species of bacteria and fungi develop primarily in damaged potato tubers during storage, including *Erwinia carotovora* subsp. *carotovora* (the cause of soft or wet rot), *Pythium ultimum* (causing watery wound rot) and *Phoma exigua* var. *foveata* (causing gangrene). *Penicillium digitatum*, which causes green mould of citrus fruits, commonly enters through wounds caused during picking, grading and packing. Seemingly insignificant abrasions of the plant surface may be most significant in disease development. An important pathogen of oilseed rape, *Sclerotinia sclerotiorum* (the cause of stem rot), enters stems via decaying petal tissue or wounds. *Botrytis cinerea*, which causes grey mould on oilseed rape, also requires rotting petals or other damage before infection can occur.

Wounding caused by pests can provide entry sites for pathogens. There may be a quite specific relationship between an insect pest species and a bacterial or viral pathogen, as already described in the section on dispersal. Conversely, the insect or nematode pest may, by wounding the plant, simply provide an entry site for a soil-borne pathogen, for example.

The final category of wound agents comprises the pathogens themselves. Damage caused by a primary pathogen, for example *Phytophthora infestans* (causing late blight) on potato tubers, frequently allows entry by one, or, in the case of the potato tuber, several, secondary pathogens, including *Erwinia carotovora* subsp. *carotovora*, the cause of soft rot of potato tubers.

Colonisation

After a pathogen has obtained entry to the plant, the next stage is for it to advance into the plant tissue. There are many ways by which this colonisation occurs, and the mode of nutrition of the specific pathogen is important in this respect. The objective of a biotroph or obligate parasite is to obtain nutrients from its host without killing it. To this end, colonisation by the biotroph must cause minimal disruption to the host plant. Specialised feeding hyphae called **haustoria** (Figure 3.1) may be produced which develop in intimate contact with the host plasmalemma yet do not kill it. Colonisation of internal tissues by biotrophs is likely to be restricted. Epidermal cells, for example, are the main tissues colonised by powdery mildews. There is evidence for pectolytic and cellulolytic enzyme production by some biotrophs in order to effect penetration and colonisation, but it is considered that such potentially disruptive enzymes are produced only locally and in small amounts. Conversely, colonisation by necrotrophs may not be so subtle. Although haustoria may be produced in certain host/pathogen combinations, many necrotrophs also produce enzymes which separate and break down plant cells, and toxins which directly kill cells. Colonisation of tissues by necrotrophs is frequently more extensive than colonisation by biotrophs, but varies in extent, from localised leaf spots to an almost total colonisation of xylem tissue by vascular pathogens. A few pathogens, notably *Phytophthora infestans*, initially exhibit a biotrophic mode of nutrition and later adopt the necrotrophic mode. *Phytophthora infestans* forms haustoria in host cells at the beginning of its colonisation phase but quickly kills cells and feeds on the dead plant material.

Infection and colonisation by fungi

In order to demonstrate the diversity of methods by which fungi infect and colonise plants, a few selected examples will be described.

Infection of lucerne by soil-borne Fusarium pathogens

The diseases caused by *Fusarium* spp. are a vascular wilt which is caused by *F. oxysporum* f.sp. *medicaginis* and a crown and root rot caused by *F. avenaceum*, *F. culmorum* and other species. This example illustrates the diversity in infection and colonisation that can occur even within one genus of plant pathogenic fungus. The influence of root exudates on resting propagules of *Fusarium* spp. is variable. However, some spores germinate near to or on the root surface of lucerne (Figure 3.2i). Appressoria may be formed, but the fungi usually penetrate tissue by growing between plant cells (Figure 3.2ii). *Fusarium avenaceum* and *F. culmorum* then extensively colonise young root tissue, mainly intracellularly (Figure 3.2iii), whereas the more specialised vascular pathogen, *F. oxysporum* f.sp. *medicaginis* grows through tissue mainly intercellularly and its growth is directed towards xylem vessels which are then colonised (Figure 3.2iv) (Parry & Pegg, 1985). Such an extensive fungal growth of *F. oxysporum* f.sp. *medicaginis* in xylem vessels is typical of other vascular wilt fungi including *Verticillium albo-atrum*. Physical blockage of the xylem and the possible production of toxins by the fungi result in the typical wilt symptoms observed in

Figure 3.2. Infection and colonisation of lucerne by *Fusarium* spp. (i) Macroconidium of *F. culmorum* germinating on lucerne root surface, (ii) penetration of root by hyphal growth between cells, (iii) intracellular colonisation of young root tissue by *F. avenaceum* and *F. culmorum*, (iv) extensive colonisation of mature xylem vessel by *F. oxysporum* f. sp. *medicaginis*, (v) microconidium of *F. oxysporum* f. sp. *medicaginis* in xylem vessel. H, hypha; Mi, microconidia.

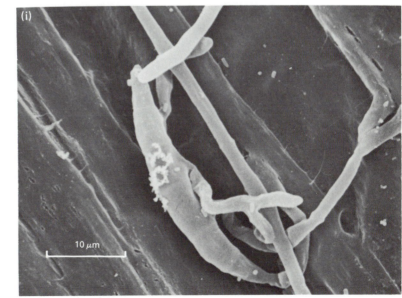

Fig. 3.2.(ii)

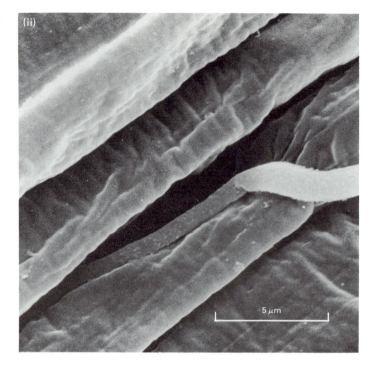

Fig. 3.2. (iii)

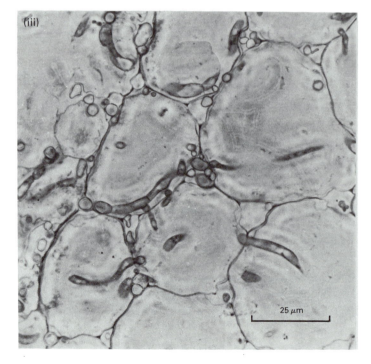

infected plants. The rapid colonisation of considerable lengths of xylem in relatively short periods by vascular wilt fungi has been attributed to the production of microconidia in the xylem (Figure 3.2v). These small spores are taken up in the transpiration stream; when they reach the perforation plate at the end of a xylem vessel they germinate and grow through the plate, producing more microconidia at the other side. Using this method of xylem colonisation, *Fusarium oxysporum* f.sp. *cubense*, the cause of Panama disease of bananas, can migrate from the bottom to the top of an 8 m tall banana tree in under 2 weeks.

Fig. 3.2(iv)

5 μm

Infection of wheat by Puccinia graminis *f.sp.* tritici, *the cause of black stem rust*

Resting spores (teliospores) of this biotrophic fungus require periods of alternate freezing and thawing before they will germinate in the spring. The basidiospores produced on germination are carried in air currents to the alternate host, the barberry plant. Here the basidiospores germinate and penetrate the epidermal cells directly. The fungus then grows mostly intercellularly and haustoria are formed in host cells. After a relatively short period (3–4 days), spermogonia emerge through the host epidermis, which produce minute spores called spermatia. When spermatia are transferred, for example by insects, to other compatible spermogonia, fertilisation takes place and aeciospores are formed in aecia on the barberry leaves. Aeciospores can infect susceptible wheat plants by means of germ-tubes penetrating stems and leaves via stomata. The resulting mycelium first grows intercellularly then aggregates towards the wheat epidermal tissue where it forms a mat of mycelium just below the epidermis. Uredospores are formed on the mat, which bursts

Fig. 3.2.(v)

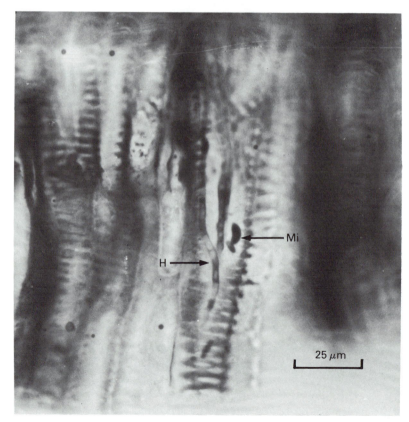

through the epidermis. Uredospores are wind dispersed and can re-infect healthy wheat tissue if relative humidity is sufficiently high for spore germination. The new mycelium grows intercellularly, haustoria are formed, and within 10 days, more uredospores are produced. Later in the season, the uredia produce teliospores, or telia may develop from recent uredospore infections. As with most biotrophs, there is little tissue degradation as a result of infection and colonisation. However, there can be significant yield losses during a severe rust epidemic. A mass of rust pustules develops, covering the photosynthetic area and disrupting the epidermis, causing increased water loss. More importantly, photosynthates are used by the fungus rather than being translocated to fill the grain.

Infection and colonisation by bacteria

Bacteria cannot directly penetrate host plant surfaces. As already stated, bacteria enter plants through wounds or natural openings such as points of lateral root emergence, stomata, lenticels and secretory glands (hydathodes and necarthodes). A water continuum between outer and inner surfaces of plants aids entry via natural openings. Tissue damage during storms also aids entry of bacteria. Wind-driven rain can induce local water soaking of plant tissue which is conducive to bacterial entry and spread. Wind and hail damage may expose xylem vessels, providing direct access to stems. Once bacteria have gained access to a plant, colonisation occurs intercellularly through natural spaces or in the vascular tissue. Many bacteria produce pectic enzymes which catalyse the breakdown of pectic substances joining cells, facilitating intercellular colonisation.

Once inside the plant, bacteria continue to multiply and, in the absence of pectic enzymes, the physical pressure of this bacterial growth may enlarge the intercellular spaces. Toxin production may also increase invasive ability, e.g. in subspecies of *Pseudomonas syringae* which cause blights and leaf spots of many crops. An example of a bacterium producing pectic enzymes is *Erwinia carotovora* subsp. *atroseptica*, which causes the disease blackleg in potato haulm tissue and soft rot in potato tubers, arguably the most important disease of stored potatoes. Bacteria are washed down infected stems onto stolons and daughter tubers and become most concentrated initially at the 'heel' end, where the stolon meets the tuber. Alternatively, tubers may become contaminated at lifting or during washing. Entry is through wounds or lenticels and bruising leads to more severe rots than cuts, possibly because it induces tissue water soaking. Wet potatoes or potatoes stored in high humidities are more vulnerable to rotting. The presence of water may result in enlarged lenticels which facilitate entry by bacteria, and oxygen depletion of underlying tissue, leading to cell leakage and further water soaking. These conditions favour colonisation of cortical tissue and the low oxygen tension stimulates pectic enzyme synthesis and secretion. Soft-rotting *Erwinia* species produce a number of pectic enzymes, the most important of which is endo-polygalacturonide (pectic) lyase. These enzymes are distributed in plant tissue in advance of the bacteria, causing maceration. The separated cells die, releasing water and nutrients, and bacterial invasion continues. Under conditions of high temperature and humidity and poor ventilation (lower oxygen tension), *E. carotovora* subsp. *carotovora* can liquefy a potato tuber in a few days.

Infection and colonisation by viruses

Infection, or transmission, of viruses is usually effected by vectors (Chapter 2). Colonisation of plants by viruses is not well understood. Intracellular movement can occur via plasmodesmata, protoplasmic bridges connecting adjacent cells. This movement is usually slow and depends upon the rate of circulatory movement of the protoplasm within the cell, the rate of production of infective units within the cell and the number of cell exits provided by the plasmodesmata. The virus is understood to be carried in plasmodesmata by diffusion and protoplasmic streaming. There is no evidence to suggest that viruses can pass through cell walls by diffusion. More rapid and longer distance movement of viruses around plants occurs in the vascular system, usually in the sieve tube elements of the phloem. Viruses frequently move towards growing regions in plants and the spread of a virus in the phloem may follow a similar pattern to that of any assimilate transported in the phloem. In some systemic virus diseases, e.g. potato leaf roll, viruses appear to be restricted to the phloem and a few adjacent parenchyma cells. In systemic viral infections caused by many of the mosaic viruses, e.g. tobacco mosaic virus, all parenchymatous cells of the plant are invaded. Some viruses appear unable to infect certain parts of plants, for example apical meristems. This is put to good effect in the vegetative propagation of certain crops, for example virus-free potato plants are cultured from plant meristem tips (virus tested stem cuttings).

Epidemiology

If infection and colonisation of plants by pathogens was a rare occurrence, perhaps happening only on a few individual plants during any one growing season, disease would have little or no economic significance. Clearly this is not the case and one of the consequences of infection and colonisation of individual plants is the eventual production of more infective propagules. Pathogen spores produced on agricultural crops can frequently find new hosts nearby and, if weather conditions are suitable for infection, a disease epidemic may develop.

A disease **epidemic** can be defined as a progressive increase in the incidence of a particular disease within a defined population of host plants over a time-scale related to the maturation of the crop (Dickinson & Lucas, 1982*a*). Epidemiology is the study of the factors which lead to this increase of disease. If we consider the interactions between the host, pathogen and environment and disease development in individual plants, as discussed in Chapter 1, the same interactions and factors will determine the initiation and progress of a disease epidemic in a population of individuals, i.e. a crop. The main reason that disease epidemics are common in agriculture and rare in nature is that farmers grow crops in which individual plants are close to each other and often genetically identical. Indeed, varieties of the major arable crops, including cereals in the UK, can only be registered if they are distinct from other varieties, uniform and stable over two or more years' field trials. Such 'clones' are all likely to have a similar degree of resistance or, more importantly, susceptibility to a variety of pathogens or different races of the same pathogen. Consequently, if a source of virulent inoculum is introduced into the crop and weather conditions are optimal for growth and spread of the pathogen, there is a high risk of a severe disease epidemic occurring in the absence of any control measures. The progress of disease epidemics has been described mathematically by Vanderplank (1963), the pioneer in this area. No survey of plant disease epidemiology would be complete without some consideration of his ideas. In addition, an understanding of a mathematical model of epidemiology may lead to a better grasp of how disease epidemics can be halted, i.e. disease control.

Simple and compound interest diseases

By studying many papers on individual plant diseases, Vanderplank recognised an essential difference in the ability of pathogens to spread over time. **Simple-interest** diseases, which can be compared with money accumulating in a bank account giving simple interest are dependent upon the amount of inoculum (or money) in the crop (account) at the beginning of the season (financial year). Simple-interest diseases are **monocyclic**, i.e. only one cycle of infection occurs during a growing season. Most are soil-borne and initial inoculum consists of resting spores or propagules in soil or on stubble or debris. Epidemics of pathogens with simple-interest disease epidemiologies progress comparatively slowly but can nonetheless be highly significant, particularly in perennial crops where pathogen populations may build-up over a number of years or in continuous cultivation of crops in the same field. Vanderplank also considered that simple-interest diseases involved pathogens which had a 'low birth rate' (comparatively few spores are produced) but that this was compensated for by their 'low death rate' (some of the spores produced may remain dormant in the soil). Examples of pathogens which exhibit a simple-interest disease epidemiology are *Gaeumannomyces graminis* (causing take-all of cereals), *Plasmodiophora brassicae* (the cause of club root of brassicas), *Fusarium oxysporum* f.sp. *cubense* (the cause of Panama disease of bananas) and *Verticillium dahliae* (which causes *Verticillium* wilt of cotton). If the progress of a typical simple-interest disease is plotted against time, a straight line may be produced which shows an increase of infected individuals or percentage infection over a number of months or years (Figure 3.3).

Figure 3.3. A simple-interest disease epidemic: the progress of infection of cotton plants with *Verticillium dahliae* (the cause of wilt). (After Ashworth *et al.*, 1979.)

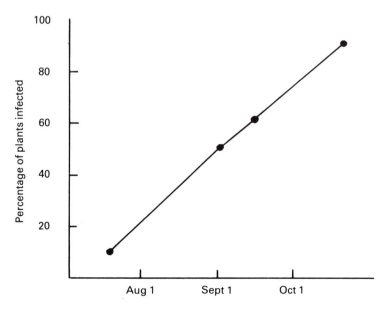

In contrast to simple-interest diseases, Vanderplank suggested that some diseases increased at '**compound interest**' during the early stages of an epidemic. In such cases, as in the case of a bank account which generates interest upon interest, pathogens can reproduce a number of times in a growing season (**polycyclic**), resulting in an exponential build-up of disease in a relatively short time. This does not imply that the rate of increase of disease, *r*, always stays constant in compound-interest diseases. Just as the financial rate of interest varies as a result of the changing financial climate, so the infection rate may vary as, for example, the weather varies. However, it is most convenient to use average values of *r* in the calculations. Vanderplank considered that compound-interest diseases involved pathogens which had a high birth rate (many spores produced), but also a high death rate (short-lived spores and much wastage). Pathogens with compound-interest disease epidemiologies include *Phytophthora infestans* (the cause of late blight of potatoes), *Erysiphe graminis* (which causes cereal powdery mildew) and *Puccinia striiformis* (causing yellow rust of cereals). A typical untransformed compound-interest disease curve gives the sigmoid graph characteristic of population growth in a limited environment. Figure 3.4 shows the compound-interest disease curve for a natural epidemic of *P. infestans* which occurred in a crop

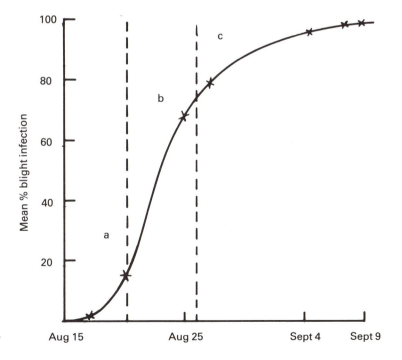

Figure 3.4. A compound-interest disease epidemic: progress curve of late blight of potatoes (*Phytophthora infestans*) in a crop of King Edward in the Midlands, UK, 1987.

of King Edward potatoes on a farm in the Midlands (UK) during 1987.

The three phases of the sigmoid curve are:

(a) the initial lag phase due primarily to the small amount of pathogen inoculum present;

(b) the exponential phase when neither the inoculum nor the amount of host tissue is limiting;

(c) the decline phase due largely to the lack of host tissue left to colonise.

Mathematical description of disease epidemics

Vanderplank (1963) suggested that a pattern of disease epidemic development as represented in Figure 3.4 could, at least in its early stages, be mathematically described by the following equation:

$$x = x_0 e^{rt} \tag{1}$$

where x = the proportion of disease at any one time, x_0 = the amount of initial inoculum, r = average infection rate, t = time during which infection has occurred, e = base of natural logs.

One constraint on this equation concerns disease at high levels. Logarithmically increasing disease requires infections to be independent. Once disease exceeds a relatively low level, for example 5%, lesions or pustules begin to overlap and are not independent: the amount of healthy tissue is effectively limiting logarithmic disease development. However, at early stages of a disease epidemic ($x \leqslant 5\%$ or, by convention $x \leqslant 0.05$), r can be derived from Equation 1 by taking logarithms and transposing:

$$\log_e x = \log_e x_0 + rt \tag{2}$$

$$rt = \log_e x - \log_e x_0 \tag{3}$$

$$r = \frac{1}{t} \log_e \frac{x}{x_0} \tag{4}$$

For practical use of Equation 4,

let amount of disease present at time $t_1 = x_1$

let amount of disease present at time $t_2 = x_2$

Equation 4 then becomes:

$$r = \frac{1}{t_2 - t_1} \log_e \frac{x_2}{x_1} \tag{5}$$

This equation can now be used to calculate r when levels of disease do not exceed 5%. As previously described, above 5% infection the progress of the epidemic is dependent upon the amount of plant tissue left to colonise and the factor $(1-x)$ is introduced into Equation 5 giving

$$r = \frac{1}{t_2 - t_1} \log_e \left[\frac{x_2 (1 - x_1)}{x_1 (1 - x_2)} \right] \qquad (6)$$

The product of this equation, r, is sometimes referred to as the **apparent infection rate**

Calculations of r, the logarithmic infection rate (Equation 4), or r, the apparent infection rate as in Equation 6, may be useful in determination of the efficacy of control measures such as resistant varieties or fungicides in the field. The more effective the control measure, the more r will be reduced. It may also be useful to compare values of r for different diseases of the same host or a variety of host/pathogen combinations.

Example of calculation of r (apparent infection rate)

The percentage of leaf area affected by mildew in a fungicide-untreated crop of barley was 10% after 45 days and 20% after 55 days. In an adjacent fungicide-treated crop, there was 7% mildew at 45 days and 12% at 55 days.

In the unsprayed plot, $x_1 = 0.10$, $x_2 = 0.20$, $t_2 - t_1 = 10$.
Substituting these values into Equation 6

$$r = \frac{1}{10} \log_e \left[\frac{0.2 (1 - 0.1)}{0.1 (1 - 0.2)} \right]$$

$$r = \frac{1}{10} [0.81]$$

$$r = 0.081 \text{ units per day (unsprayed).}$$

In the fungicide sprayed plot, $x_1 = 0.07$, $x_2 = 0.12$, $t_2 - t_1 = 10$.

$$r = \frac{1}{10} \log_e \left[\frac{0.12 (1 - 0.07)}{0.07 (1 - 0.12)} \right]$$

$$r = \frac{1}{10} [0.59]$$

$$r = 0.059 \text{ units per day (unsprayed).}$$

As expected in this example, the progress of the disease epidemic was slowed down as a result of fungicide application.

The examples given in Table 3.1 show interesting points with regard to the nature and control of disease epidemics. The progress of a disease epidemic is reduced dramatically by the use of resistant varieties. In addition, the compound-interest diseases (potato blight, yellow rust and tobacco mosaic virus) have higher apparent infection rates than the simple-interest diseases (wilt of bananas and cotton), demonstrating a quicker build-up of disease in the crop. However, some simple interest diseases, e.g. *Verticillium* wilt of cotton, can, under optimum conditions with a sufficiently high

Table 3.1. *Examples of apparent infection rate (r)*

Pathogen (disease)	Host		*r* (units/day)
Phytophthora infestans (late blight)	Potato	Susceptible variety	0.42
		Moderately susceptible variety	0.16
Puccinia striiformis (yellow rust)	Barley	Susceptible variety	0.27
	Wheat	Moderately susceptible variety	*c.* 0.10
Tobacco mosaic virus	Tobacco		*c.* 0.10
Fusarium oxysporum f.sp. *cubense* (wilt – Panama disease)	Banana		≤0.0014
Verticillium dahliae (wilt)	Cotton	Susceptible variety	*c.* 0.05 (Figure 3.3)

(After Zadoks & Schein, 1979.)

initial density of inoculum, build-up almost as quickly as some compound-interest diseases.

Further reading

Fry, W. E. (1982). *Principles of Plant Disease Management*. New York: Academic Press.

Huang, Jeng-sheng (1986). Ultrastructure of bacterial penetration in plants. *Annual Review of Phytopathology*, **24**, 141–57.

Vanderplank, J. E. (1963). *Plant Diseases: Epidemics and Control*. New York: Academic Press.

Wheeler, B. E. J. (1976). *Diseases in Crops. Institute of Biology Studies in Biology No. 64*. London: Arnold.

Wood, R. K. S. (1967). *Physiological Plant Pathology*. Oxford: Blackwell.

Zadoks, J. C. & Schein, R. D. (1979). *Epidemiology and Plant Disease Management*. Oxford: Oxford University Press.

4 What effect does disease have on the crop?

Introduction

The effects of disease on a crop may or may not be obvious. *Disease damage*, the first section of this chapter, may be visibly manifested as symptoms such as discolorations, abnormal growth, rots and physiological wilts. Less apparent damage from disturbance of the physiology of the plant will also be discussed. Any consideration of disease damage must include some means of measuring or assessing levels of disease and, as far as possible, relate this assessment to yield loss. The second and third sections of this chapter will therefore be concerned with *disease assessment* and *crop loss*.

Disease damage: symptoms

Discolorations

These include **chlorosis**, which is yellowing of plant tissue as a result of the breakdown of chlorophyll, and **necrosis**, which often follows chlorosis and is a browning or blackening of cells as they die. Lesions on plants frequently consist of a central area of necrotic tissue, often the initial site of fungal infection, surrounded by a chlorotic halo, e.g. *Alternaria brassicae*, the cause of target spot of oilseed rape (Figure 4.1).

Abnormal growth

This occurs as a result of hormonal disturbances in the host plant by the invading pathogen. The plant may be stimulated to produce more cells (**hyperplasia**), bigger cells (**hypertrophy**) or a combination of the two. The net result may be the enlargement of a plant organ, resulting in a gall or a club. Probably the best known example of this is clubroot of brassicas caused by *Plasmodiophora brassicae* (Figure 4.2). Other abnormal growth patterns include the formation of new meristems, the stimulation of adventitious roots, or the loss of apical dominance leading to the development of side shoots. One example of an abnormal growth pattern is that which occurs in rhizomania ('root madness') disease of sugar beet. The main tap root becomes bearded by a massive proliferation of lateral rootlets (Figure 4.3). Included in the category of abnormal growth could be the formation of scabs (new cork cambia), as occurs in both common and powdery scab of potatoes (Figure 4.4). Perhaps the most common abnormal growth pattern is that of stunting. Many viruses and root-infecting fungi cause reduced growth of plants. Fewer leaves and flowers result in less fruit and grain.

Figure 4.1. Target spot of oilseed rape caused by *Alternaria brassicae*. (Schering Agriculture.)

Figure 4.2. Clubroot of oilseed rape caused by *Plasmodiophora brassicae.* (DAFS.)

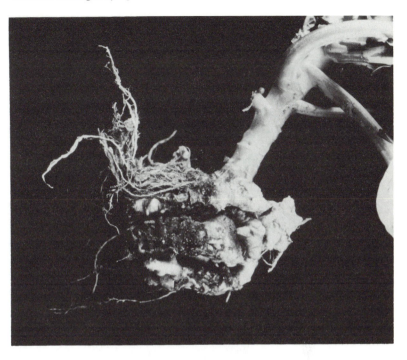

Figure 4.3. Rhizomania disease of sugar beet caused by beet necrotic yellow vein virus. (M. J. C. Asher, Brooms Barn Experimental Station.)

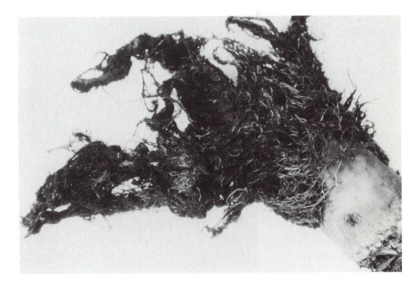

Rots

A rot is simply the disintegration of tissue. There are two types of rot; a soft or wet rot, most often caused by bacteria; and a dry rot, most frequently attributable to fungi. Rots are common in plant organs which have homogenous parenchymatous tissue, e.g. fruits and tubers. Potato tubers can suffer both from dry and wet rot in store. A common and significant problem in potato stores is that of soft rot caused by the bacterium *Erwinia carotovora* subsp. *carotovora* (Figure 4.5). Dry rot in potato tubers is caused by

Figure 4.4. Common scab of potatoes caused by *Streptomyces scabies*. (Plant Breeding International Cambridge.)

Figure 4.5. Soft or wet rot of potato tubers caused by either *Erwinia carotovora* subsp. *atroseptica* or *E. carotovora* subsp. *carotovora*. (Purdue University.)

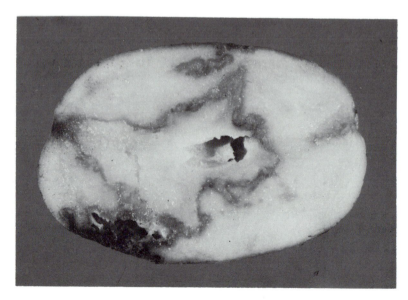

Fusarium spp. Rots can also occur in roots and stems. Generally they are more restricted than those in tubers, but nevertheless can be troublesome. *Fusarium* species are also responsible for root and stem base rots of many arable crops including cereals and legumes.

Physiological wilts

Vascular wilt pathogens such as *Fusarium oxysporum* formae speciales and *Verticillium albo-atrum* colonise the vascular tissue of many plants and cause a characteristic wilt symptom. The wilting is not only a consequence of the physical blockage of xylem vessels but also a result of toxin production by the fungi. Some bacterial species also cause wilts, e.g. *Erwinia stewartii* (bacterial or Stewart's wilt of maize).

Disease damage: physiological

All the main physiological processes in the plant can be disturbed by the invasion of a pathogen, athough it should be remembered that other non-biological factors such as stress, and physical and chemical damage may lead to similar disturbances.

Photosynthesis

At a gross morphological level, photosynthesis can be dramatically reduced by pathogens growing over or killing areas of green leaf and stem tissue. There is also evidence that pathogen invasion results in changes to the photosynthetic process itself. These effects may, however, be less deleterious than expected when net rates of photosynthesis in both infected and uninfected tissue are considered. Generally, there is a significant and compensatory increase in the rate of photosynthesis in the healthy tissue in diseased leaves. This rate continues to rise until the final stages of disease, when senescence and death of the leaf occur. In the case of necrotrophic pathogens such as *Rhynchosporium secalis* on barley, toxin (rhynchosporiside) produced may directly affect photosynthesis. Rhynchosporiside may cause leaf stomata to open and thereby increase photosynthesis in healthy parts of affected leaves.

Senescence of leaf tissue in the final stages of disease may be delayed in the areas surrounding certain pathogens. The 'green island' effect is perhaps most commonly observed in cereals colonised by the brown rust pathogen (*Puccinia hordei*). A green ring of tissue in a senescent leaf surrounds individual rust pustules. It is almost as if the biotroph is trying to keep the plant alive for as long as possible to supply its own needs. There is indeed evidence to suggest that photosynthesis is occurring normally in green island tissue. However, the formation of green islands may not be a deliberate attempt by the pathogen to keep itself alive, but rather a natural response of the plant to the presence of a site of active metabolism; a nutrient 'sink'. Green islands occur in senescing tissue which has virtually lost its usefulness as a photosynthesising organ and is metabolically relatively inactive. The presence of a biotrophic, metabolically active pathogen (nutrient sink) may simply attract nutrients to a leaf which under normal circumstances would die.

Habeshaw (1984) argues that the changes which have been frequently observed in photosynthesis in infected plants may not primarily result from the active involvement of the pathogen. Rather, plants respond in a normal way to damage caused by the pathogen and not to the presence of the pathogen itself.

Respiration

One physiological change which commonly occurs in plants after pathogen invasion is that the respiration rate increases. Some of the early work with powdery mildews and rusts showed that respiration rate increased initially, just before the development of visible symptoms. As disease developed, respiration rate continued to rise and usually reached a maximum of two to four times that in uninfected tissue at sporulation. By mechanical removal of fungal mycelium and measurements of mycelial respiration, it was determined that the major rise in respiration was in the host tissue itself, and not simply the additional respiration of the pathogen. It is clear, therefore, that the host metabolism is stimulated by certain pathogens. This increase in respiration can, at least in part, be responsible for yield losses in diseased plants. The biochemical basis for this stimulation is uncertain but there are essentially three hypotheses: the pathogen uncouples respiration and/or the pathogen changes respiratory patterns and/or the plant diverts photosynthates to secondary metabolism and protective processes.

Firstly, the process of oxidative phosphorylation (formation of ATP in the electron transport system) may be inhibited either directly or indirectly by pathogens. Inhibition of this process results in an increase in the concentration of ADP and an increase in oxygen uptake. There is evidence that the mode of nutrition of plant pathogens is significant here. Some necrotrophs, e.g. *Pyrenophora teres*, the cause of net blotch of barley, produce toxins which may directly uncouple oxidative phosphorylation or affect membrane permeability. Disturbance of mitochondrial membranes can cause ion leakage and a concomitant increase in respiration. Smedegaard-

Petersen (1984) observed that respiration increase in susceptible barley leaves inoculated with *P.teres* started within 24 h, reached a maximum after 7 days and then decreased rapidly as leaves died. In contrast, in susceptible barley leaves inoculated with the biotrophic fungus *Erysiphe graminis* f.sp. *hordei*, the increase in respiration was not observed until 3 days after inoculation. The respiration rate remained high until sporulation ceased and senescence occurred (Figure 4.6). Smedegaard-Petersen concluded from these data that, in the case of the biotrophic fungus, the increase in respiration was not directly part of pathogenesis, but rather a result of increased biosynthesis in the plant: the host plant was providing extra nutrients to the biotroph and, in addition, the pathogen may have been acting as a nutrient sink.

A change in respiratory patterns, which involve biochemical pathways usually absent or much less common in healthy plants, is the second hypothesis to account for stimulation of respiration in infected plants. There is some evidence, particularly in the case of biotrophic fungi, for a shift from the normal glycolytic pathway resulting in the production of pyruvic acid to a 'side-line', the

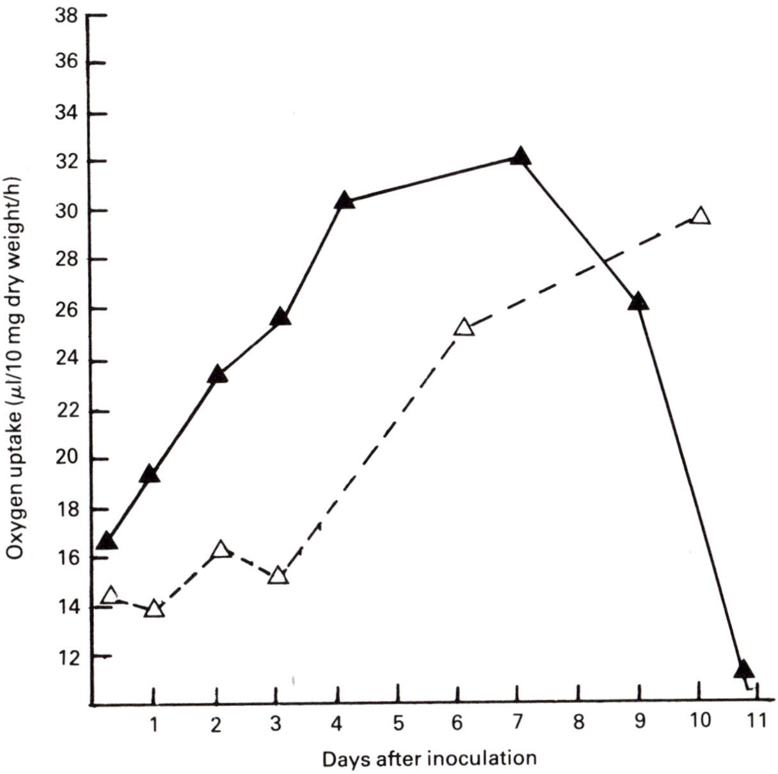

Figure 4.6. Time course of respiration in two susceptible barley cultivars inoculated with, respectively, the net blotch pathogen *Pyrenophora teres* (a necrotroph) and the powdery mildew pathogen *Erysiphe graminis* f.sp. *hordei* (a biotroph). ▲——▲, inoculated with *P. teres*; △——△ inoculated with *E. graminis*. (After Smedegaard-Petersen, 1984.)

pentose-phosphate pathway. An increase in this pathway results in higher concentrations of NADPH being produced. NADPH is the main source of reducing power in plant biosynthetic processes and more of the compound would be required to supply a biotrophic fungus with nutrients.

So far, increases in respiration have been discussed in terms of compatible host/pathogen combinations. However, there is evidence that respiration increases in resistant plants when challenged with avirulent pathogen isolates. Active defence mechanisms in plants, such as the production of lignin and phytoalexins (Chapter 5), are associated with increased biological activities. Plants have to expend energy to produce these compounds. Although plants may not show any visible disease symptoms when challenged by an avirulent pathogen, in the case of barley as a host and the powdery mildew pathogen, Smedegaard-Petersen & Tolstrup (1985) demonstrated a significant increase in respiration in the incompatible combination. The net result of this wasted energy was a 7% reduction in grain yield in a resistant variety. Further to this, these workers suggest that 'non-significant' leaf surface saprophytes such as *Cladosporium* elicit similar active, energy-consuming defence reactions. This, they propose, may be part of the explanation why yields have frequently been observed to increase after fungicide application in the absence of any disease symptoms.

Transport systems

Transport systems in plants may be conveniently divided into those which operate over short distances, i.e. in and out of cells, and those which operate over long distances, i.e. via the xylem and phloem. According to Farrar (1984), transport systems can be affected by plant pathogens in three ways: physical damage; changes in mass transfer (alterations in water and nutrient distribution); and non-nutritional chemical effects including those caused by plant growth regulators, phytoalexins and toxins.

In long-distance transport, there are many reports of a rise in the concentration of nutrients in infected leaves. This may be a result of increased transpiration into a leaf which has suffered a ruptured cuticle as a result of pathogen attack. There is also evidence that exports of nutrients in the phloem are reduced in some host/pathogen combinations. The mechanisms by which this reduced export occurs are uncertain. Indeed the mechanism of phloem transport in healthy plants is still largely unresolved. It is possible, however, that the pathogen causes damage to phloem tissue, reduces the supply of carbohydrates in the phloem, or affects phloem loading and transport in some way. The main function of mature green tissue is the production of nutrients during photosynthesis and their export to areas of the plant (sinks) where they are most needed, such as growing points. If as a result of pathogen invasion this export is not occurring properly, yield reductions may be expected.

Pathogens may also affect short-distance transport of nutrients within an organ, for example a leaf. Nutrients frequently accumulate around infection sites, as previously discussed in the sections on photosynthesis and respiration. Biotrophic pathogens may simply be acting as nutrient 'sinks', thus creating a concentration gradient along which nutrients can diffuse passively. If the concentration gradient does not favour passive uptake, it is possible that the pathogen may actively transport nutrients. In the colonisation of tissue by necrotrophic fungi in particular, there is evidence of disruption of membranes by physical penetration of fungal hyphae, enzymatic degradation, or metabolic disturbances of membrane properties by toxins. Short-distance transport in some or all of these cases may be aided by nutrients becoming available as a result of 'leaky' membranes.

Disease assessment

In order for decisions to be made regarding the most cost-effective control strategies in a particular situation, an initial assessment of the level of disease should be made. Disease assessments are normally carried out with the help of disease keys which are most often specific to pathogen–host combinations. Disease assessment keys are fairly easy to devise; however, the most successful keys will be objective, simple and quick to use. Objectivity is important in that the key should hold good for different observers in different areas and different seasons. It is important, therefore, to have clear diagrams and descriptions. Terms like 'severe' disease are useless unless they are qualified. In order to devise a suitable disease assessment key, the progress of disease must first be closely studied in the field from sowing to harvest. A study of the morphology and development of healthy plants may also provide useful back-up information. A picture key or descriptive key can then be devised for the pathogen/host combination. Examples of picture keys for two crops are shown in Figures 4.7 and 4.8. The basis for picture keys is a fairly accurate representation of the plant part to be assessed, for example, a cereal leaf. Life-like disease symptoms such as pathogen pustules or necrotic areas are then drawn onto the

Figure 4.7. Picture key for
assessment of powdery
mildew (*Erysiphe graminis*
f.sp. *hordei*) on barley.
(After MAFF, 1976a. ©
Crown copyright 1989.)
Notes on assessment
To assess leaf diseases
on cereals estimations
should be made of the leaf
area affected by each of
the diseases present
on appropriate leaves.
For recording:
(a) disease incidence up
to the start of jointing,
assess infection on the
lowest green leaf.
(b) disease development
and early differences
in susceptibility from
jointing to flowering,
assess infection on
leaf 3*.
(c) infection likely to be
related to loss in
yield, assess infection
at the milky-ripe
growth stage (usually
10–14 days after
flowering) on **leaf 2**
and **flag leaf**.
ALWAYS NOTE
GROWTH STAGE AND
LEAF RECORDED.
It is suggested that
samples for assessment at
or after jointing should be:
for plots: 10 fertile tillers
selected at random
for fields: 25 fertile tillers
selected at random
Score individual leaves for
percentage area affected

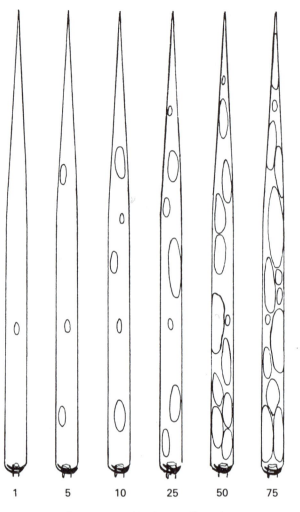

Percentage of leaf area affected

1   5   10   25   50   75

plant background at a convenient range of disease levels. The precise area that these symptoms occupy can be calculated quite easily by drawing the key on graph paper or by using an area meter. Samples of diseased material from the field can now be compared with levels of disease on the picture key and quite accurate disease assessments made. Descriptive keys, as shown in Tables 4.1 and 4.2 relate a disease severity score or percentage area affected to a description. A number of both descriptive and picture keys are presented in publications of the Ministry of Agriculture, Fisheries and Food (1976a,b) and National Institute of Agricultural Botany (1985). Logarithmic scales may be used in both types of key. The

by disease: assess lesion area excluding any chlorosis for all diseases except rusts where associated chlorosis should be added to the rust area. Interpolate between the levels depicted if necessary. It is also useful to record the area of dead leaf tissue not apparently associated with any specific disease or green leaf area. Assess the upper surface only unless the lower surface is more severely affected, in which case assess the lower surface only.

* At jointing leaf 3 arises from the second lowest node in wheat and the third lowest node on barley; from flag leaf emergence it is the third from the top of the plant.

reason for this is the logarithmic build-up of disease in certain pathogens. It is probably as important epidemiologically to detect differences in disease at low levels, for example 1–10% infection, as it is to detect differences between 10 and 100%. Fungicides are more effective when applied at low disease levels.

Notes on assessment which should accompany all keys will have information on the optimum growth stage for carrying out the assessment and recommendations for sampling. Growth stage keys are available for many crops. At each assessment the growth stage must be defined, and occasionally a different disease key is used depending on the growth stage. Method of sampling and sample size are of obvious importance. Usually samples are taken at random; sample size depends on the aims of the investigation. Recommendations for sample size are based on a compromise between the statistical ideal and what is practical in the field. For example, cereal trials carried out in small plots demand a minimum of 10 samples per plot and trials in larger fields require a minimum of 50 samples (Figure 4.7). The assessment method used depends on the purpose of the assessment, as there is a hierarchy of precision. When ranking plots, e.g. for comparisons of efficacy of fungicides or varieties,

Table 4.1. *Descriptive key for assessment of eyespot of wheat* (Pseudocercosporella herpotrichoides)

Infection category	Disease severity description
0	Uninfected
1	*Slight eyespot* (one or more small lesions occupying less than half the circumference of the stem)
2	*Moderate eyespot* (one or more lesions occupying at least half the circumference of the stem)
3	*Severe eyespot* (stem completely girdled with lesions; tissue softened so that lodging would readily occur)

Notes on assessment
1. Examine 20 tillers per 20 m² plot
2. Assign each tiller to one of the infection categories above.
3. Write the number of tillers in each category on the record sheet.
4. An index will be calculated from the data as follows:

$$\text{Disease index} = \frac{(0 \times a) + (1 \times b) + (2 \times c) + (3 \times d)}{(a + b + c + d)} \times \frac{100}{3}$$

where a, b, c and d are the number of tillers examined which fall into the categories 0, 1, 2, and 3, respectively.
(After Scott & Hollins, 1974.)

Figure 4.8. Picture key for assessment of leaf spot (*Ascochyta fabae*) on pods of field and broad beans. (From MAFF, 1976b. © Crown copyright 1989.)

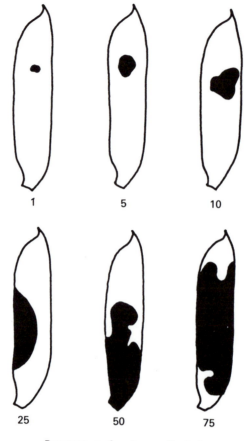

Percentage of pod area affected

Notes on assessment
Carry out the assessment at growth stage 8.1 just before the pods turn brown.
For field assessments select 10 sites at random along any one diagonal. Sample 5 stems from each of the 10 sites; choose the 1st, 3rd, 5th, 7th and 9th stems from the starting point of the samples.
For each stem:
Count the total number of pods, and the number infected with *Ascochyta*.
Estimate the % area infection of each of the infected pods using the key above as a guide. For each pod, assess only the most heavily infected side and use this figure as the actual % area infection of the whole pod. If this method proves too time-consuming, record the number of pods on each stem which fall into the following categories:
0% *Ascochyta*
<25% *Ascochyta*
>25% *Ascochyta*
For plot assessments it is suggested that similar counts should be made on 10 plants selected at random from each plot.

whole-plot assessments can be used. Yield loss studies require a percentage area assessment key. Epidemiological studies, e.g. correlation of environmental factors with infection, may necessitate precise daily counts of lesions or numbers of pustules.

It is important to maintain standards between assessors. This can be achieved by each individual assessing the same plots, comparing results and making any necessary adjustments.

Crop loss

In control strategies, disease assessments are most useful if they can be related to crop losses. Plant pathogens cause crop losses in a variety of ways, some more obvious than others. There may be a direct yield reduction as a result of disease or a reduction in quality. The latter may still be expensive for the farmer in terms of downgrading of produce. It is easy to overlook the cost of control

Table 4.2. *Descriptive key for assessment of foliage blight of potatoes* (Phytophthora infestans)

Blight (%)	Disease severity description
0	Not seen on field
0.1	Only a few plants affected here and there; up to 1 or 2 spots in 12 yards radius
1	Up to 10 spots per plant, or general light spotting
5	About 50 spots per plant or up to 1 leaflet in 10 attacked
25	Nearly every leaflet with lesions, plants still retaining normal form: field may smell of blight, but looks green although every plant is affected
50	Every plant affected and about half of leaf area destroyed by blight; field looks green flecked with brown
75	About ¾ of leaf area destroyed by blight: field looks neither predominantly brown nor green. In some varieties the youngest leaves escape infection so that green is more conspicuous than in varieties like King Edward, which commonly shows severe shoot infection
95	Only a few leaves left green, but stems green
100	All leaves dead, stems dead or dying

Notes on assessment
In the earlier stages of a blight epidemic parts of the field sometimes show more advanced decay than the rest and this is often associated with the primary foci of the disease. Records may then be made as, say 1 + pf 25, where pf 25 means 25% in the area of the primary foci.

Make successive assessments at intervals of 7–14 days to record progress of blight. Begin in good time as both nil and starting date records (0.1%) are important. Difficulties in judging allowance to be made for stem blight on a particular date are overcome by making another assessment later.

(From Anon., 1947.)

measures to reduce disease and the inevitable crop losses as a result of spraying machinery damage. When disease results in significant yield loss, an additional area of crop may need to be grown in order to satisfy demand. In effect, a proportion of the land is unproductive. This concept was developed by Ordish (1952) and is known as 'untaken harvest'. Crop losses in some instances may be occurring undetected. This is particularly true of root-infecting pathogens which are largely facultative parasites and may cause chronic but mild diseases of roots. Virus diseases of plants may remain undetected for years until new technology emerges to free the plant from the virus and the yield suddenly increases. Yield increases in fungicide-treated 'healthy' plants may, at least in part, be a result of the control of mild, undetected pathogens.

A number of techniques have been developed to relate disease intensity to yield loss. Tarr (1972) divides these methods into statistical and experimental.

Statistical methods of measuring crop losses involve the collection and analysis of reports of disease incidence, estimated crop losses and yield figures. Obviously the greater the body of data accumulated, the more significant it becomes, but results from such studies must be treated with caution. One method involves the collection of data on crop yields in years of differing disease incidence. The major weakness of these data is that many other factors besides disease can influence yield. The weather, crop variety, crop husbandry and pests may all contribute to annual yield fluctuations. However, if one major disease can be identified as responsible for a fluctuation in yield, this method may be useful. Experienced agronomists may be able to make reasonably accurate forecasts of yield quite early in the season. If there is a difference between expected and observed yields, which can be attributed to a major disease attack, this can become the basis for a statistical method of estimating yield loss. Yield data collected before and after the introduction of a disease control measure may also provide useful information relating yield loss to disease. For example, the introduction of a curly-top resistant variety of sugar beet in the USA raised yields from 12.5 to 35 t/ha. However, this increase in yield could in part be due to other factors such as differences in the innate yielding capacity of the two varieties. In some countries, the use of questionnaires sent to farmers and agricultural officers can provide useful information on crop losses. The validity of these data largely depends on the competence of the recorders.

A more accurate relationship between disease incidence and crop loss can be derived using **experimental methods**. The first of these to be discussed involves yield comparisons between infected and uninfected plants or between plants suffering from different disease severities. Disease can be built up by artificial inoculation, or naturally infected plants can be used. Experiments may be done on a large field scale, on smaller experimental plots or even on individual plants. Experimental variables, such as other diseases, pests and variations in soils, should be minimised. A very popular development of this technique involves monitoring disease on single tillers of cereals (Richardson, Jacks & Smith, 1975). The main advantage of single tiller assessment is the minimum resources required. Large numbers of single tillers are labelled and levels of disease, as a result of either natural or artificial infection, are assessed. Disease assessments may be repeated through the season.

At harvest, the labelled ears are individually threshed and yield components such as number of grains, 1000-grain weight and yield per ear are then taken. Regressions of yield parameters can then be determined and a very useful relationship of $x\%$ disease $= y\%$ yield loss may be derived. This method has formed the basis for many of the yield loss formulae for cereal diseases (Table 4.3). Loss estimates from such equations are unlikely to be totally accurate, but they give an approximate idea of the relative significance of cereal diseases and hence suggest where fungicide activity should be directed.

The next experimental method for assessing yield loss involves comparisons between resistant and susceptible varieties. The most obvious problem with this method is any unknown difference in the innate yielding capacity between varieties. The technique is most useful, therefore, if varieties are closely related. The different components of a 'multiline' variety, which only differ in their race-specific resistance to a pathogen, may be effectively used in this context (see Chapter 5, Disease resistance).

Finally, comparisons can be made between plots of infected plants and plots kept free from disease by fungicide applications. This method is probably the most widely used and a very accurate way of relating disease levels and yield loss. Experiments are frequently carried out on fairly small plots, for example 20 m × 2 m, or they may be on a larger field scale. Fungicides are used as a tool to free the treated crops from the constraint of disease. As such, they are applied on more than one occasion according to a spray programme which may be uneconomic for the farmer to adopt. In addition, the fungicides are often 'broad spectrum', aimed at killing all diseases, therefore it is difficult in an average year to relate any yield increase in treated plots to control of one specific disease. However, fungicides may be selected which have specific activity against a single disease; alternatively, broad-spectrum fungicides may be used at sites where it is known that a single disease predominates. It has to be borne in mind that fungicides frequently increase yield in the absence of obvious disease. A small percentage of the yield increase as a result of fungicide application may, therefore, be attributable to this.

The previously described experimental methods for assessing yield losses as a result of disease are suitable for relatively small experiments over restricted areas of crop. In some instances it may be useful to relate disease to yield loss over a larger area, for example a whole country. The results of a national disease survey of cereal diseases in the UK are reported by Cook & King (1984).

Table 4.3. *Formulae used in yield loss estimates for cereal diseases*

Disease	Relationship
Spring barley	
Mildew	$y = 2.5 \sqrt{x_i}$
Brown rust	$y = 0.4 x_{ii}$
Rhynchosporium	$y = 0.5 x_{ii}$
Winter wheat	
Mildew	$y = 2.0 \sqrt{x_i}$
Septoria	$y = x_{iii}$
Yellow rust	$y = 0.4 x_{iii}$
Eyespot	$y = 0.1 x_m + 0.36 x_s$
Sharp eyespot	$y = 0.05 x_m + 0.26 x_s$

$y = \%$ loss in grain yield,
$x_i = \%$ disease on leaf 3 at GS 58,
$x_{ii} = \%$ disease on leaf 2 at GS 75,
$x_{iii} = \%$ disease on flag leaf at GS 75,
$x_m = \%$ tillers with moderate symptoms at GS 75,
$x_s = \%$ tillers with severe symptoms at GS 75.
(From Cook & King, 1984.)

Disease levels in a very large number of randomly selected crops were estimated by members of the Agricultural Development and Advisory Service (ADAS). These disease level data were then translated into estimates of yield losses by using the formulae in Table 4.3. The average total loss caused by all diseases was estimated to be 10.3% per year (1967–80) for spring barley and 5.2% per year (1970–82) for winter wheat. After a thorough consideration of the cost of fungicide application, balanced against the value of the lost grain as a result of disease, Cook & King (1984) concluded that the application of fungicides to cereals is associated with a high probability of a net financial benefit. However, as a result of the survey, these workers also reported that there was considerable scope for improving the identification of crops which are most at risk from disease and therefore potentially most responsive to fungicides.

Yield losses in cereals have been the primary consideration so far. Losses are manifested in the main by reduced grain yields. There is an increasing body of evidence, however, that certain cereal diseases may also reduce cereal grain quality, particularly 1000-grain weight (Hims, 1987). The effect of diseases on the quality of vegetables and fruit is more dramatic than on quality of cereals. In vegetables, even small blemishes diminish the value of a crop by affecting its visual acceptability. Fruit and vegetables damaged by disease will be downgraded by the market inspectorate and farmers' profits dramatically reduced. It is common for there to be a 33% difference in the market value of class I and class II crops. European Community (EC) common quality standards are set for the majority of field vegetables and fruit. There are certain minimum requirements for all classes of produce. Generally produce has to be sound, firm, and with no signs of rotting, softening or blemish which would detract from its edibility and storage. In some crops, for example shelling peas, the minimum requirements specifically state that produce must be free from damage caused by insects and diseases. Crops are then graded into at most a further four classes; 'extra', I, II and, where applicable, III. To meet class I standard, it is common that crops should be disease-free. Although EC quality standards are high, standards of the large supermarket buyers are often more exacting. If there are plentiful supplies of produce, supermarket buyers can be very selective and even the slightest blemish can result in crop rejection. Modern packaging of produce frequently involves the use of transparent plastics, for example to cover pre-washed vegetables. Any blemish on the surface is therefore easily detected and likely to result in rejection by the consumer. In addition, polythene packs without ventilation holes punched in them are ideal incubators for most diseases and it is very risky to package and store material this way if it harbours pathogens.

Further reading

Ministry of Agriculture, Fisheries and Food (1976). *Manual of Plant Growth Stages and Disease Assessment Keys*. Pinner, Middlesex: MAFF Publications.

National Institute of Agricultural Botany (1985). *Disease Assessment Manual for Crop Variety Trials*. Cambridge: NIAB publication.

Tarr, S. A. J. (1972). *The Principles of Plant Pathology*. London: Macmillan.

Wood, R. K. S. & Jellis, G. J. (1984) (ed.). *Plant Diseases, Infection, Damage and Loss*. Oxford: Blackwell.

5 How do we control disease?

Introduction

The object of any method of control is to stop the build-up of disease in a crop. The progress of a disease epidemic has already been described mathematically (Chapter 3):

$$x = x_0 e^{rt}$$

The object of control is to reduce x, the amount of disease, to as low a value as possible; ideally $x = 0$. In order to do this, we need to influence the variables on the other side of the equation. There are several ways to reduce x_0 (initial inoculum) to zero. Any planting material must be as free as possible from pathogen contamination. Growing certified seed and using fungicide seed treatments will help here. Legislative control measures are also useful in excluding diseases not already present in a region or a country. Inoculum in the environment surrounding the crop can be reduced by various methods of cultural control. For example, soil-borne inoculum can be reduced by good crop rotations and correct disposal of plant debris. Volunteers and weeds are often good initial sources of inoculum which should be eliminated as far as possible.

The rate of progress of the epidemic, r, will be considered next. Even if the attempts at reducing x_0 have been successful, fresh inoculum may arrive during the growing season which can initiate a disease epidemic. Again, a number of methods are available to reduce r. Growing resistant or tolerant varieties will slow down the progress of a disease epidemic. The use of fungicides which may kill the fungus or slow down its rate of growth is also very important. Biological control, in the few cases where it can be applied practically, will also slow down the disease epidemic, but is perhaps not as effective in the short term as resistant varieties or fungicides.

Finally t, the timing of the disease epidemic, can also be influenced by control measures. With regard to cultural control, crops should be planted at an optimum time. Early and late-sown crops often suffer with more disease. Fungicides can also be used to influence the duration of a disease epidemic. A 'systemic' seed treatment, which is taken up by the young plants, may protect them from airborne inoculum and thereby delay the onset of a disease epidemic. Foliar-applied 'protectant' fungicide treatments act in a similar way.

An increasingly popular approach to disease control is to combine and balance all the available methods into an 'Integrated Control' programme. Such a strategy is an attempt to influence all the variables discussed above (x_0, r and t).

In practice, control measures must be cost-effective for the farmer and easy to implement. Ideally they must not be phytotoxic, toxic to the operator, or pollutants. It is immediately obvious that some control methods pose more problems than others in meeting these critera. Plant pathology is applied most directly to agriculture in the control of plant diseases and therefore the principles behind each of the following major methods of control will be described in some detail: legislative or regulatory control; cultural control; disease resistance; chemical control; biological control; forecasting disease; and integrated control.

Legislative control

Legislative or regulatory control is an attempt to exclude pathogens from areas where they do not already exist. Legislative control may operate on a national or international scale. The basis of any disease legislation involves prohibition or restriction of the introduction of plants or plant parts, which is backed up by some form of inspection. National legislation takes a number of forms. Eradication of alternate hosts may be backed up by legislation. For example, eradication of barberry, the alternate host of *Puccinia graminis* (the cause of black stem rust of wheat) was made law in the 18th century in British colonial settlements in North America. More frequently and more recently, national plant disease legislation affects the farmer directly when governments have to be informed about outbreaks of indigenous but geographically localised diseases. So-called **notifiable diseases** in the UK include red core disease of strawberries caused by *Phytophthora fragariae*, fire blight disease of apples and pears (*Erwinia amylovora*), in some areas, wart disease of potatoes (*Synchytrium endobioticum*), plum pox (sharka disease) of plums (plum pox virus) and progressive wilt of hops (*Verticillium albo-atrum*). Under their respective Orders or Acts, such diseases must be reported; diseased material must not be transported or sold and may have to be destroyed. In addition, non-indigenous diseases, such as rhizomania (caused by beet necrotic yellow vein virus) on sugar beet, bacterial blight (*Pseudomonas syringae* pv. *pisi*) on peas and brown rot (*Pseudomonas solanacearum*) and ring rot (*Corynebacterium sepedonicum*) on potatoes, must also be reported in the UK. Outbreaks of such potentially serious non-indigenous diseases are dealt with much more rigorously, involving quarantine measures. Crop destruction, disinfection of equipment, sterilisation of soil and extra physical barriers around infected fields were used in an attempt to contain the first outbreak of rhizomania in the UK during 1987. In addition, legislation has been introduced specifically to restrict the spread of rhizomania in the UK. The 'Disposal of Waste Order 1988' is an attempt to control the soil-borne disease by regulating the disposal of soil and debris from imported potatoes and root vegetables. The Order requires the registration of processors and hauliers of waste from these processors, and the approval and licensing of tips where the waste is dumped.

Another important aspect of regulatory control in the domestic situation involves seed certification schemes. In the UK it is illegal to sell seed of any major arable crop unless it has been certified as meeting certain specified minimum standards of quality, including freedom from disease. Such schemes exist for cereals (wheat, barley, oats and rye), beet, oil and fibre plants and cruciferous fodder plants, vegetables, field beans and field peas, grasses and herbage legumes, and potatoes. There are also schemes for horticultural crops such as strawberries and raspberries, together with ornamental trees. Crops are inspected by trained inspectors both in the field and after harvest. Some crops may be grown specifically to meet standards higher than those required for ordinary certified seed and there may be a gradation in quality. For example, potatoes grown for seed may be classified as super elite, elite, AA or CC in the field according to eligibility criteria such as geographical location, isolation from other crops and rotation, together with quality criteria such as number of rogues and levels of specified diseases (Table 5.1). A further confirmatory inspection of the tubers, where disease and pest levels are assessed, is necessary prior to marketing (Table 5.2). Such high standards of seed health are essential in potato growing as there are many important diseases which are seed-borne.

Table 5.1. *Standards for potato seed production in the growing crop*

General The crop must be vigorous and fit for inspection, must not have suffered an undue amount of roguing, must not be affected with any disease or pest rendering it unsuitable for seed purposes and must not exceed the tolerances for rogues, diseases etc. set out in the table below (percentage plants inspected).

| Disease and defects | VTSC[a] | Scheme | | | |
		Super elite %	Elite %	AA %	CC %
Rogues, undesirable variations, wildings, bolters or semi-bolters	Nil	0.05	0.05	0.1	0.5 (of which not more than 0.2% shall be rogues)
Leaf roll	Nil	0.01			
Severe mosaic disease	Nil	Nil	0.1	0.25	2.0
Tobacco veinal necrosis virus (TVNV)	Nil	Nil			
Mild mosaic disease	Nil	0.05	0.5	1.0	5.0
Blackleg	Nil	0.25	0.5	1.0	2.0

[a]VTSC = virus tested stem cuttings.

Notes
1. Certificates are issued by the Agricultural Departments of the UK and seed is classified as Basic (grades VTSC, Super elite, Elite and AA) which can be multiplied further or Certified (grade CC) which must not be multiplied further.
2. VTSC and Super elite grades can only be produced in 'protected' areas, i.e. regions of the UK where the risk of aphid-borne virus infection is lowest. Such areas include Scotland, Northern Ireland, parts of Northumberland and Cumbria.
3. Generation limits are as follows: VTSC 3 years; Super elite 3 years; Elite 3 years; AA no limit; CC only used for production of ware.

(From Ministry of Agriculture, Fisheries and Food (MAFF) UK seed potato classification (certification) scheme.)

The increase in transport of plant material across international boundaries, particularly evident after the Second World War, has resulted in international co-operation in attempts to reduce disease, especially by plant disease legislation. The United Nations Food and Agriculture Organisation (FAO) organised in 1951 an International Plant Protection Convention with the aim 'of securing common and effective action to prevent the introduction and spread of pests and diseases of plants and plant products'. By the early 1980s, 81 countries had become members of the world-wide convention, which is now regionally organised. For example, in Europe and the Mediterranean, the organisational body is the European and Mediterranean Plant Protection Organisation

Table 5.2. *Standards for all grades of potato seed tubers (other than virus tested stem cuttings) prior to marketing (percentage by weight)*

Specified disease (pathogen), or pests, damage and defects	Individual tolerances	Group tolerances	Collective group tolerances
Group I (a) Wart disease (*Synchytrium endobioticum*) (b) Potato tuber eelworm (*Ditylenchus destructor*) (c) Potato cyst eelworm (*Heterodera* species infesting potatoes) (d) Ring rot (*Corynebacterium sepedonicum*) (e) Brown rot (*Pseudomonas solanacearum*) (f) Potato tuber moth (*Phthorimaea* = *Gnorimoschema opercullella*) (g) Potato spindle tuber viroid (h) Colorado beetle (*Leptinotarsa decemlineata*) *Note:* Items (d) to (h) are not known to occur in Great Britain	Nil		
Group II Blight (*Phytophthora infestans*) Blackleg (*Erwinia carotovora* var. *atroseptica*) Soft rots including Watery wound rot (*Pythium ultimum*), Pink rot (*Phytophthora erythroseptica*) and Pit Rot Dry Rot (*Fusarium* species) Gangrene (*Phoma exigua* var. *foveata*) Frost-damaged tubers	1%	1%	
Group III Skin spot (*Oospora* = *Polyscytalum pustulans*)	2%		
Group IV Black scurf (*Rhizoctonia solani*)	3%	4%	
Common scab (*Streptomyces scabies*)	4%		5%
Powdery scab (*Spongospora subterranea*)	4%		
Group V External blemishes or tubers other than diseased tubers whose shape is atypical for the variety	3%	3%	
Group VI Dirt or other extraneous matter	1%	1%	1%

(From MAFF seed potato classification (certification) scheme.)

(EPPO), which regularly produces bulletins including information on newly identified pests and pathogens and phytosanitary regulations. Most member countries have their own disease legislation with regard to imported plant material. Regulations are often divided into three sections. Firstly, a general section may prescribe the form of health certificate to accompany any imported material. It will lay down rules for inspection and disposal of material if contaminated. Penalties for violation of regulations may also be specified. Secondly, there will be a list of prohibited imports, and finally, a list of restrictions of imports of material from specified areas. The Plant Health Order 1987 is the prescribed legislation which gives information on, among other things, import restrictions into the UK.

Implementation of plant disease legislation requires some form of policing. In the UK this is normally carried out by Customs and Excise officers. Documentation, including any phytosanitary certificates, is either checked at ports or inland inspectors examine consignments and certificates at their destination. With exports, consignments of plants are usually inspected and certified on the nursery of production. In addition to this, spot checks are carried out by officers of the Plant Health and Seed Inspectorate (PHSI) on material both entering and leaving the country. Particular caution is taken over material imported from a country with disease problems not indigenous to the UK. New outbreaks of disease throughout the world are carefully monitored by the PHSI and restrictions on import of plant material may be imposed. Suspect material may be put into quarantine for a period of time in order to detect pathogens which are difficult to diagnose in dormant material. In the UK, it is mainly suspect horticultural produce that has to undergo quarantine. This is generally undertaken in commercial nurseries under strict supervision. Countries which have more problems with undetectable diseases such as viruses may have specific quarantine stations for this purpose.

Cultural control

In the broadest sense, cultural control of plant diseases encompasses all aspects of crop husbandry which influence disease development, including resistant varieties and biological control. However, the latter subjects are covered separately and this section will deal specifically with the influences of crop rotation, cultivation, crop hygiene, eradication of alternative hosts and sowing on disease development.

Crop rotation

An important aspect of the agricultural revolution of the 18th century was the introduction of crop rotations. Fear of disease or other unknown catastrophe was probably the general principle behind the rotation practised at the advent of scientifically based agriculture. The availability of crop protection chemicals has, comparatively recently, reduced the significance of crop rotation as a means of disease control, but for certain soil- and trash-borne diseases, rotation is still the cheapest and most effective method for reducing disease. The efficacy of rotation in reducing soil-borne plant pathogens will depend largely on the ability of specific pathogens to survive in the soil (Chapter 1). For example the take-all pathogen of cereals, *Gaeumannomyces graminis*, can be effectively reduced by just a one or two year break from susceptible crops, which include wheat, barley and rhizomatous grass weeds. Conversely, resting spores of *Plasmodiophora brassicae* (clubroot of brassicas) can survive for decades in soil. A further consideration in the planning of rotations on the farm should include proximity of early and later sown crops. For example, early-sown potatoes should not be planted in fields adjacent to main-crop potatoes. *Phytophthora infestans* (blight) may infect early potatoes before their harvest and thus provide a good source of inoculum for adjacent main-crop potatoes where the problem would be much more severe. Similarly, winter and spring barley should not be sown in adjacent fields, in order to reduce spread of such diseases as mildew and leaf blotch. An alternative to crop rotation is the practice of monocropping on a specific field until disease becomes a problem. The field is then used to grow another crop which is not susceptible to the disease. For example, onions may be monocropped until white rot (caused by *Sclerotium cepivorum*) becomes damaging and then a non-susceptible crop such as a cereal could be grown and the onions moved to another 'clean' field.

Cultivation

Although few farmers would specifically adopt a cultivation method because of its ability to reduce disease, there is some evidence that recent trends in methods of cultivation are affecting disease build-up, particularly in cereals and soybeans. In the USA, continuous soybean production, together with reduced cultivation (no ploughing), has resulted in the upsurge of a number of diseases including *Phytophthora* root and stem rot, brown stem rot, stem canker and brown spot (Chapter 11). In maize, the incidence of grey leaf spot and anthracnose has increased for similar reasons (Chapter 12).

When a farmer grows two successive cereal crops the first problem to be overcome before he can cultivate the soil is disposal of cereal straw and stubble. The main options are chopping and incorporation into the topsoil, deep-ploughing, or burning. Burning straw and stubble is probably the best way of reducing trash and stubble-borne foliar pathogens such as *Septoria* spp., net blotch and *Rhynchosporium* on cereals. This also applies to oilseed rape, peas and beans and any other crops where straw and stubble remain after harvest. However, public opinion and byelaws are often vehemently opposed to burning, and straw incorporation and ploughing are now the most popular methods of disposal. One would expect an increase in trash and stubble-borne pathogens as a result, particularly using the incorporation method, but there are limited data to support this. However, results of recent trials by the Agricultural Development and Advisory Service (ADAS), reported by Yarham (1986), suggest that take-all levels are highest in straw incorporated treatments when compared with straw burning. This may be a consequence of the effects of cultivation on soil structure, rather than a direct effect on take-all inoculum.

Decisions on the method of straw incorporation used may be closely linked to subsequent cultivation techniques. Farmers can either invert the soil to a pre-determined depth by a shearing and twisting action using a mould-board plough before cultivation, deeply cultivate, directly drill seeds into an undisturbed soil or practise reduced cultivation where soil is disturbed to a depth just sufficient to produce a seedbed. Again, one would expect more disease problems with reduced cultivation and direct drilling. In practice, there is little evidence for this in the UK. Results of a comprehensive survey of take-all after mould-board ploughing, reduced cultivation and direct drilling showed no significant difference in percentage of plants infected with take-all (Yarham & Norton, 1981). Indeed, there is some evidence with another stubble-borne disease, eyespot, that direct drilling actually reduces the disease (Table 5.3). This unexpected result may be a consequence of a change in plant growth habit after direct drilling. Precision-drilled plants are frequently more prostrate, with their crowns nearer to the soil surface, than conventionally drilled plants. Perhaps there is some added protection of the stem base from splash-borne eyespot spores in more prostrate plants. It is clear that the effect of cultivation method on disease is complex and frequently obscured by other factors such as crop growth and the environment.

Table 5.3. *Effects of cultivation methods on incidence of eyespot at Rawreth, Essex*

Mean percentage of plants with eyespot (spring)			Mean percentage of shoots with eyespot (summer)		
Mould-board[a] plough	Reduced[b] cultivation	Direct[c] drill	Mould-board plough	Reduced cultivation	Direct drill
45	40	27	49	41	30

[a] *A mould-board plough* is used in the UK to invert the soil to a pre-determined depth by a shearing and twisting action.
[b] *Reduced cultivation* is any system which reduces the power requirement for cultivation.
[c] *Direct drilling* involves sowing seeds into an undisturbed soil together with residues of the previous crop.

(From Yarham & Norton, 1981.)

Crop hygiene

Clean, tidy farms and fields are less likely to have disease problems than farms with piles of discarded crops and trash. Crop hygiene includes correct disposal of straw and stubble, as described previously, and also correct disposal of unwanted crops such as discarded potatoes. One of the best sources of potato blight inoculum is a pile of sprouted discarded potatoes close to a potato field. Crop hygiene also includes eliminating 'volunteers' or 'groundkeepers'. These are self-sown plants which can act as overwintering hosts for many pathogens, particularly the biotrophs which need green living material to survive on. In a few cases, shed seed may be infected before it reaches the ground. For example *Alternaria* infection of oilseed rape pods results in dark lesions and premature ripening, splitting of pods and shedding of seed. A proportion of the shed seed will be infected with *Alternaria* and will give rise to infected volunteers. Cereal volunteers are important as 'green bridges' for diseases such as mildew where they provide the most important means of overwintering for the biotrophic mildew pathogen. Suitable herbicides, ploughing and burning are all effective means of reducing volunteers.

Crop hygiene also includes the practice of 'roguing' a crop. This is the removal and destruction of plants in a growing crop when they become diseased. Weeds may also be rogued out of a crop. Although this process is highly labour-intensive, it may be worthwhile in valuable crops where the disease is conspicuous and slow-spreading. Roguing is sometimes practised in potato crops grown for 'seed' where a limit is placed on the number of diseased individuals in a standing crop, e.g. leaf roll of potatoes caused by the potato leaf roll virus.

Eradication of alternative hosts

Alternative hosts for fungi, bacteria and viruses have been discussed in Chapter 1. Many biotrophic pathogens can survive the winter by infecting alternative hosts which may or may not be related to the host crop plant. For some pathogens, particularly those with a well understood life cycle and narrow host range, it may be possible to reduce disease by reducing or preferably totally eliminating alternative host species. This is frequently a costly, lengthy process and may be impossible in practice. For example, barley yellow dwarf virus of cereals can overwinter on grass. It would be sensible, therefore, to eliminate grass weeds from in and around a cereal crop. However, it is impossible to kill all the grass over large areas. One of the most famous campaigns to control a plant disease involves eradication of an alternative host. *Puccinia graminis* f.sp. *tritici*, the cause of black stem rust of wheat, completes its life cycle on an alternate host, the barberry bush (see Chapter 1). In the USA a programme of barberry eradication began in 1918 and in 1942, after the destruction of around 300 million bushes, 60% of the wheat-growing areas were free of barberry. This very costly programme was only partially successful, probably because total elimination of barberry was impossible and also because uredospores can be wind-blown over long distances. However, the programme still continues today, albeit on a more limited scale. Elimination of barberry in the vicinity of wheat fields is an obvious sensible precaution.

Sowing

Date of sowing can influence disease development, particularly in autumn-sown cereal crops. Generally, although earlier-sown crops suffer with more disease problems there has been a trend towards early sowing of cereals. Between 1976 and 1979, about 5% of winter wheat crops were sown in September in the UK; the figure was nearer 20% between 1985 and 1987 (Yarham & Giltrap, 1989). Yield benefits of early sowing in the absence of severe disease are well understood. However, the higher incidence of diseases such as take-all, eyespot, sharp eyespot, mildew, *Rhynchosporium*, net blotch and *Septoria* leaf spot may offset the beneficial effects of early sowing (Table 5.4). Early-sown crops of winter oilseed rape are also prone to more disease problems. Dark leaf and pod spot (caused by *Alternaria* spp.) and powdery mildew (caused by

Table 5.4. *Effect of sowing date on take-all of winter wheat at Rosemaund Experimental Husbandry Farm, Hereford and Worcester, UK*

| Sowing date | Take-all index | | Yield (t/ha) |
	Winter	At harvest	
1983 2nd wheat			
15 September 1982	68.2	91.6	2.64 ⎱ high disease
15 October 1982	22.1	72.6	5.05 ⎰ pressure
1984 2nd wheat			
23 September 1983	14.6	22.0	10.57 ⎱ low disease
11 October 1983	0.6	17.0	10.61 ⎰ pressure

(From Simkin, Nicholson & Clare, 1985.)

Erysiphe cruciferarum) are more severe in crops sown before 20 August (MAFF, 1984). Late sowing of crops may also favour certain diseases. Late-sown winter cereals are likely to be slow to emerge and therefore suffer more from damping-off and seedling blights caused by *Pythium* and *Fusarium* spp. Later in the season, backward crops will have a profusion of young, mildew-susceptible leaves which may suffer severe mildew attacks and generate a large amount of inoculum. Optimum sowing times vary considerably according to many factors: when the previous crop was harvested; rotation of crops; geographical location; soil type; desired market; and variety. Generally, for maximum yields in the Midlands (UK), winter wheat should be sown in the first 2 weeks of October, winter barley at the end of September to the beginning of October, spring barley in the last week of February and winter oilseed rape at the end of August.

Other aspects of sowing which can affect disease development are quality of seedbed, depth of sowing and seed rate. In general a well prepared firm seedbed will allow plants to develop good root systems and become healthy and vigorous. This is particularly important in the control of take-all of cereals, which is more severe in light, 'puffy' soils. If seed is sown too deeply, emergence will be delayed, giving problems with disease similar to those of late sowing. Also, it is important to sow at optimum seed rates because too dense a stand of crop results in increased humidity within the stand, which encourages many diseases. In addition, if the distance between individual plants is reduced, aerial spores and mycelium of root-infecting fungi can spread more quickly within a crop.

Disease resistance

Introduction

The use of disease-resistant varieties is one of the pillars of modern agriculture. Indeed in some instances, resistant varieties are the only means of controlling plant disease. Much emphasis is now placed on reducing the variable costs of farming. Fungicide inputs are coming under close scrutiny, particularly as a result of the development of fungicide-insensitive or resistant pathogen isolates. The use of resistant varieties is a very common part of most 'Integrated Control' strategies.

The monetary value of growing resistant varieties is a subject which has received little attention. However, it is clear from a study by Priestley & Bayles (1988) that resistant varieties have saved, and continue to save, the British farmer money. The value of disease resistance in the popular cereal varieties grown when compared with the most susceptible cereal varieties was estimated to be £262 million per annum (1983 prices).

Before embarking on a study of disease resistance, it is worth defining some of the terms commonly used. A **resistant** plant is one which possesses qualities which hinder the development of a given pathogen. Clearly resistance involves some interaction between specific host plants and specific pathogens. This is an important concept which will be developed later in the context of the gene-for-gene hypothesis (p. 98). Conversely, a **susceptible** plant is one which is subject to infection. An **immune** plant is one which is exempt from infection. Immunity is an absolute property expressed against pathogens which cannot cause disease in a particular host. For example, cereals are immune to the potato blight pathogen *Phytophthora infestans*. Finally, **tolerant** plants are able to endure colonisation by a pathogen without showing severe symptoms of disease, or to compensate for the effects of disease.

The majority of this section on disease resistance is concerned with genetic resistance to plant pathogens. However, disease escape is considered briefly first.

Disease escape

One of the ways in which a plant can appear resistant to a disease is by a process of disease escape. Plants can escape disease by various means which are not related to the genetic resistance of the plants themselves. Some cultural methods of disease control described earlier in the chapter involve disease escape. If a crop is grown out of phase with a potential pathogen, the crop has escaped the disease. For example, spring-sown wheat escapes the diseases snow rot and snow mould, caused by *Typhula incarnata* and *Fusarium nivale*, respectively, because snow cover in spring is unusual in temperate areas.

Disease escape may be linked to some genetic factor in the host. Some barley varieties are particularly susceptible to loose smut (caused by *Ustilago nuda*). The fungus infects the barley floret during anthesis (flowering). Varieties which exhibit totally closed flowering, i.e. their anthers are not exposed to the atmosphere (or smut spores), escape loose smut infection.

Genetic resistance

Plants may exhibit **major gene resistance**, based on one or a few genes, or **polygenic resistance**, based on many genes. Genetic resistance may operate at different stages of development in a plant, **seedling resistance** working when plants are young and **adult plant resistance** when they are mature. Adult plant resistance is useful in resisting diseases under field conditions, and is sometimes also called **field resistance**. It is, at least in part, a result of the changes in structure and physiology as a plant matures. Clearly there must be a gradation between the two types of resistance and it is very difficult or impossible to specify the point at which seedling resistance becomes adult plant resistance. The distinction is often based on observations from laboratory or glasshouse experiments. Seedling resistance is generally based on major genes, whereas adult plant resistance is generally polygenic.

Major gene resistance was first discovered by Biffen in 1912. Biffen investigated the resistance of wheat to yellow rust and found that in his breeding programme, the yellow rust resistance gene behaved according to the laws of simple inheritance proposed by Mendel. The nature of polygenic resistance is much more difficult to understand and breeding for it very difficult, as will be seen later. In breeding for resistance, the easiest option has been taken. Hence major gene resistance which is relatively easy to identify and incorporate into new varieties, was sought and used extensively. Sources of new genetic resistance are usually wild relatives of the cultivated species. For example *Solanum demissum*, a wild central American species related to *S. tuberosum*, the cultivated potato, has been used over many years as a source of major genes which confer resistance to potato blight.

Gene-for-gene hypothesis

After working with major gene resistance for a number of years, it became apparent that there was some sort of relationship between specific varieties possessing major resistance genes and pathogens which could attack some varieties and not others. It was suggested that pathogens which could overcome major resistance genes in varieties must themselves have corresponding virulence genes. Hence, the **gene-for-gene hypothesis** was proposed in 1956 by Flor who worked with flax and flax rust (*Melampsora lini*). This concept proposes that in certain plant/pathogen interactions, a gene for resistance in the host corresponds to a gene for virulence in the pathogen. Since this initial proposal, the gene-for-gene hypothesis has been demonstrated or suggested in other pathogen/host combinations (Table 5.5).

The gene-for-gene hypothesis is very difficult to verify experimentally. However, in a few pathogen/host combinations, listed in Table 5.5, there is strong experimental evidence to support the hypothesis. Both of the flax rust on flax and powdery mildew on barley combinations are experimentally proven examples of the gene-for-gene hypothesis. As further research is carried out into pathogen/host relationships, it is quite possible both that the experimental proof will be found to support the hypothesis in more of these combinations, and that Table 5.5 will be extended.

An assumption that the gene-for-gene hypothesis is in operation can be of practical use. For practical purposes, the term 'factor' may be used to refer to the gene or genes involved in this relationship which have not necessarily been identified. When a new crop variety is bred, it is possible to assign it with specific resistance factors to a range of diseases. It is common, through a combination of natural mutation and selection pressure exerted on the pathogen by new resistance factors in a variety, for new races to arise in the pathogen population. These races will possess specific virulence factors capable of overcoming resistance factors in the plant. The gene-for-gene hypothesis can also be used to detect new virulence factors in the pathogen population.

The basis for determination of genetic resistance in new varieties is a historical set of **differential varieties**. These varieties have been assigned specific resistance factors (Table 5.6). One of the first interactions between pathogen and host to be understood in terms of the gene-for-gene hypothesis involved *Puccinia striiformis*, the cause of yellow rust of wheat, which will be used as an example (Table 5.7). On the basis of the resistance factors of the differential varieties, virulence factors can be assigned to races of *P. striiformis*. According to the gene-for-gene hypothesis, in order for a pathogen race to be able to infect a variety, it must have a virulence factor (gene) which corresponds to and is directed against any resistance

Table 5.5. *Pathogen/host combinations for which a gene-for-gene relationship has been shown or suggested*

Pathogen	Disease	Host
Erysiphe graminis f.sp. *tritici*	Powdery mildew	Wheat
Puccinia graminis f.sp. *tritici*	Black stem rust	Wheat
Puccinia recondita	Brown rust	Wheat
Puccinia striiformis	Yellow rust	Wheat
Ustilago tritici	Loose smut	Wheat
Tilletia caries	Bunt (stinking smut)	Wheat
Tilletia contraversa	Dwarf bunt	Wheat
Tilletia foetida	Common bunt	Wheat
Erysiphe graminis f.sp. *hordei*	Powdery mildew	Barley
Ustilago hordei	Covered smut	Barley
Puccinia graminis f.sp. *avenae*	Black stem rust	Oats
Ustilago avenae	Loose smut	Oats
Puccinia sorghi	Rust	Maize
Potato virus X	Mild mosaic	Potato
Phytophthora infestans	Late blight	Potato
Synchytrium endobioticum	Wart	Potato
Tobacco mosaic virus		Tomato
Tomato spotted wilt virus	Spotted wilt	Tomato
Fulvia fulva	Leaf mould	Tomato
Bremia lactucae	Downy mildew	Lettuce
Venturia inaequalis	Scab	Apple
Melampsora lini	Rust	Flax
Hemileia vastatrix	Rust	Coffee
Puccinia helianthi	Rust	Sunflower
Phytophthora megasperma f.sp. *glycinea*	Root and stem rot	Soybean
Xanthomonas malvacearum	Bacterial blight	Cotton

(After Vanderplank, 1984.)

Table 5.6 *Resistance factors to* Puccinia striiformis *(the cause of yellow rust of wheat) in differential varieties*

Resistance factor	Gene	Differential variety
1	1	Chinese 166
2	2	Heine V11
3	3a + 4a	Vilmorin 23
4	3b + 4b	Hybrid 46
5	5	*Triticum spelta album*
6	6	Heines Kolbein
7	7	Lee
8	8	Compair
9	9	Riebesel 47/51

(After United Kingdom Cereal Pathogen Virulence Survey (UKCPVS) Annual Reports.)

Table 5.7. *Gene-for-gene relationship in wheat varieties and wheat yellow rust (caused by Puccinia striiformis)*

		Hypothetical races of *Puccinia striiformis*										
		A	B	C	D	E	F	G	H	I	J	K
	V factors	1	2	3	4	5	6	7	8	9	1,2,3,4,6	1,2,3,4,5,6,7,8,9
Differential varieties	*R factors*											
Chinese 166	1	S	R	R	R	R	R	R	R	R	S	S
Heine V11	2	R	S	R	R	R	R	R	R	R	S	S
Vilmorin 23	3	R	R	S	R	R	R	R	R	R	S	S
Hybrid 46	4	R	R	R	S	R	R	R	R	R	S	S
Triticum spelta album	5	R	R	R	R	S	R	R	R	R	R	S
Heines Kolben	6	R	R	R	R	R	S	R	R	R	S	S
Lee	7	R	R	R	R	R	R	S	R	R	R	S
Compair	8	R	R	R	R	R	R	R	S	R	R	S
Riebesel 47/51	9	R	R	R	R	R	R	R	R	S	R	S
Varieties of wheat												
Galahad	1	S	R	R	R	R	R	R	R	R	S	S
Brock	7	R	R	R	R	R	R	S	R	R	R	S
Slejpner	9	R	R	R	R	R	R	R	R	S	R	S
Norman	2,6	R	R	R	R	R	R	R	R	R	S	S
Fenman	1,2,4	R	R	R	R	R	R	R	R	R	S	S
Longbow	1,2,6	R	R	R	R	R	R	R	R	R	S	S

R = Resistant reaction, S = Susceptible.

(After UKCPVS Annual Reports.)

factor (gene) present in the variety. A variety with a particular resistance factor R1, will be susceptible only to a pathogen with a correspoding virulence factor V1. Hence the resistance factors in any new variety of wheat can be identified. For example, Galahad is susceptible to races A, J and K, all of which possess the virulence factor V1. Galahad must therefore possess the corresponding resistance factor R1. A variety with more than one resistance factor is only susceptible to a pathogen with all the corresponding virulence factors. In practice, it is very rare to find races of a pathogen which contain only a single virulence factor (races A to I). The diversity of R factors in popular varieties favours pathogen races which have a number of V factors. Race J (V1, 2, 3, 4, 6), a typical example of the spectrum of virulence found in a yellow rust race, infects many more varieties than Race A. Any variety which possesses one or a combination of R factors, which correspond to the V factors in Race J, is susceptible to it. The logical conclusion of selection pressure on the pathogen is the development of a 'super-race' which contains V factors to match any R factor available (race K). This is also rare, possibly because such a race would contain many V factors which are unnecessary to attack existing varieties. It is worrying, however, to see how rapidly V factors build up to match new R factors. A plant breeder frequently attempts to introduce a new R factor into a variety but then, according to the gene-for-gene hypothesis, the pathogen has the capacity to match and overcome this R factor with a corresponding V factor. This happens in practice and is discussed later.

It is clear that major gene resistance involves relationships between specific host varieties and specific races of the pathogen. Another term for this type of resistance is **race-specific resistance**. Many plants possess resistance which is conferred against all races of a given pathogen, frequently referred to as adult plant or field resistance. This is normally of polygenic origin and is sometimes called **race non-specific resistance**.

Vertical and horizontal resistance
The concept of race-specific and race non-specific resistance was expanded in the 1960s by Vanderplank, who developed the idea of 'vertical' and 'horizontal' resistance. When a variety showed more or less complete resistance to some races of the pathogen but not to others, he termed this vertical (race-specific) resistance, whereas when a variety showed the same level of resistance to all races, he termed it horizontal (race non-specific) resistance. This terminology was based on the appearance of diagrammatic representa-

tions of the reactions of different varieties to a range of races of pathogen. A plant variety showing vertical resistance must have a major resistance gene, or R factor, in that it is resistant to all races of pathogen which lack the corresponding virulence gene (V factor). The amount of horizontal resistance shown by a variety does not depend on any R factor. A variety with no R factor (varieties A and B in Figure 5.1) will show the same amount of resistance to all races of pathogen, though some varieties will have a higher level of this horizontal resistance than others. A variety with vertical resistance (varieties C and D in Figure 5.1) will show complete resistance to those races of pathogen against which its vertical resistance is effective. It will also show a certain amount of resistance to those races of pathogen which can attack it; once again this level of horizontal resistance is higher in some varieties than others.

The concept of vertical and horizontal resistance to plant diseases has important implications for the plant breeder and the farmer. A variety such as variety D would be the aim: one which shows a high level of horizontal resistance as well as some vertical resistance, although variety B would also be desirable.

The 'boom and bust' cycle
As already mentioned, major gene resistance has been traditionally incorporated into new crop varieties, using fairly simple plant breeding techniques, with considerable success. At the start of its life, a new variety with major gene resistance should be totally resistant to a particular disease. Such a variety is likely to prove popular with farmers as, all other things considered, a resistant variety may give higher yields, especially in the absence of fungicides. However, in practical terms, the popularity of a variety with major gene resistance may be the prime reason for its eventual demise. The 'boom and bust' cycle, which was proposed by Priestley (1978), describes the situation, particularly with regard to cereal varieties when, after a period of widespread cultivation of a variety with major gene resistance, it succumbs to disease and its popularity declines (Figure 5.2).

A new variety with good disease resistance may gradually become more popular. However, the widespread use of such a variety favours the races of pathogen with matching virulence genes. Eventually a critical stage is reached, determined partly by the frequency of virulence genes and partly by environmental factors, resulting in a severe disease epidemic. The variety has then effectively 'bust', its popularity declines and the frequency of specific virulence genes in the pathogen population also declines.

Figure 5.1. Graphical
representation of vertical
and horizontal resistance
to pathogen races.

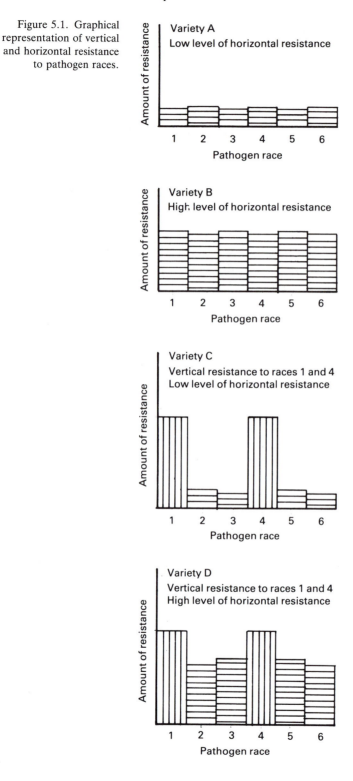

The plant breeder then effectively has to go back to the drawing board and breed another variety possessing other resistance genes. The situation has been described as a treadmill and varieties which may have taken up to 10 years to breed have, in a number of instances, 'bust' after only 1 or 2 years' commercial usage. A good example of this occurred in the winter wheat variety Stetson, bred in the UK. It first entered the government-funded testing system at the National Institute of Agricultural Botany (NIAB) in 1979. The variety was totally resistant to yellow rust and had a good level of resistance to other diseases. Consequently there was a good deal of interest in the variety and it was marketed largely on the strength of its disease resistance. Stocks of Stetson were built up and in 1982 it was launched commercially. During the summer of 1982, yellow rust was detected on the variety: a new race of the pathogen *Puccinia striiformis* had developed which overcame the variety's major gene resistance. This was the beginning of the end for Stetson and, after a further 2 years on the NIAB *Recommended List of Cereal Varieties*, it was removed.

The initial detection and virulence determination of the new yellow rust race, together with warnings for farmers, was carried out by the United Kingdom Cereal Pathogen Virulence Survey (UKCPVS). This body was set up in the late 1960s after severe and widespread yellow rust epidemics had devastated winter wheat varieties, such as Rothwell Perdix and later Joss Cambier which had hitherto been very popular. The UKCPVS monitors virulence patterns in mildew of wheat, yellow and brown rust of wheat, mildew of barley, yellow and brown rust of barley, *Rhynchosporium* and net blotch of barley and mildew and crown rust of oats. Results are published annually in the UKCPVS *Annual Report*.

Generally, as in the case of Stetson, when a variety 'busts' as a result of a new or increased virulence, its commercial life is over. However, this is not always the case. A new race of the powdery

Figure 5.2. The boom and bust cycle. (From Priestley, 1978.)

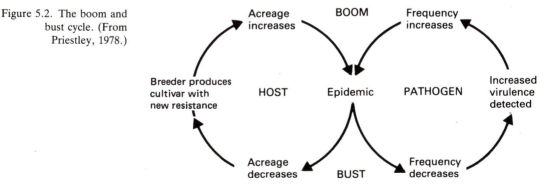

mildew pathogen, with a virulence gene which enabled it to overcome the major gene resistance in the spring barley variety Triumph, was identified as early as 1977. However, it was not until 1983 that the frequency of the gene in the pathogen population increased sufficiently to cause serious problems on this increasingly popular variety. Triumph's popularity was at its peak in 1983 when it occupied 22% of the total (winter and spring) UK barley acreage (Table 5.8). Despite the breakdown in resistance Triumph did not disappear from the fields and its decline in popularity occurred at a rate similar to that of any other barley variety without its disease problems. New, higher yielding or better quality varieties are continually being introduced, which naturally supersede older varieties. The secret of the continued popularity of Triumph was its acceptable yield coupled with its potential to be used by maltsters in the brewing industry. The extra premium attracted by 'malting' barley in many farmers' minds more than compensated for the cost of fungicide sprays.

A further, rather worrying concept with regard to the breakdown of major gene (vertical) resistance was proposed by Vanderplank (1968). The **vertifolia effect** is defined as the erosion of horizontal resistance by plant breeders when their breeding material is protected from disease by vertical resistance. Vanderplank proposed that in some instances when a variety has 'bust', the disease epidemic is particularly severe because there is little horizontal resistance remaining in the variety. The vertifolia effect is the subject of much debate, as it appears to hold true in some pathogen/host combinations, but not in others.

It is evident that major gene resistance falls short of the ideal, long-lasting resistance that both plant breeders and farmers require. Consequently, there is now an increasing emphasis on trying to breed new varieties of crop plants which possess 'durable resistance'.

Durable resistance

Major gene resistance against biotrophic pathogens such as mildew and rust is generally highly unstable and non-durable. Major gene resistance against necrotrophs or facultative parasites is often considerably more durable. Polygenic resistance, on the whole, appears also to be durable. According to Johnson (1981), varieties which remain resistant to pathogens that have highly developed variety-specific pathogenicity, even though they are extensively cultivated in environments favourable to disease, may be described as possessing durable resistance.

Table 5.8. *Popularity and resistance of Triumph spring barley to powdery mildew caused by* Erysiphe graminis

Harvest year	NIAB Recommended List resistance rating for mildew (1–9)[a]	Certified seed (t)[b]	% total barley market[b]
1980	8	4609	2
1981	8	41329	13
1982	8	58370	20
1983	8	65480	22
1984	7	49049	19
1985	2	45164	16
1986	2	36091	14
1987	2	30204	12
1988	2	28232	10

[a]A high figure indicates a high degree of resistance.
[b]From MAFF Seed Certification Scheme statistics.

Examples of major gene resistance which have stood the test of time include the 60 or more years that some varieties of cabbage have remained resistant to yellows disease caused by *Fusarium oxysporum* f.sp. *conglutinans*. Such examples are fairly rare, but when they do arise they should not be neglected by the plant breeder. Polygenic resistance in the absence of major resistance genes is the basis for durable resistance in most instances. The partial resistance of many barley varieties to brown rust caused by *Puccinia hordei* is such an example and this has been achieved simply by removing the most susceptible lines from breeding programmes. The resistance of certain wheat varieties to *Septoria nodorum* is a further example of polygenic durable resistance although in this case the level of resistance is low and could be improved upon.

Selection for polygenic, potentially durable resistance is not always as simple as in the case of barley brown rust. If the heritability of the resistance is low in the host, resistance cannot be selected for on an individual plant basis. For example, testing for resistance of potato varieties to virus diseases is often done on large plots. In any breeding programme where polygenic, durable resistance is the objective, plants which exhibit totally resistant or extremely susceptible reactions should be excluded. The aim is to breed plants with moderate levels of resistance. Such resistance is likely to be polygenic and as such stands a greater chance of being durable.

Table 5.9. *Characteristics of major gene and polygenic resistance*

	Major gene	Polygenic
Expression	Usually clear cut: expressed from seedling stage to maturity, or may be expressed in mature plants only	Variable response; not usually expressed in seedling stage; resistance increases as plant matures
Mechanism	Generally an immune or hypersensitive host reaction	Reduced rate and degree of infection, development and/ or reproduction of pathogen
Efficiency	Highly efficient and specific against certain pathogen races. May mask extreme susceptibility to other races	Variable, but operates against all races of the pathogen
Genetic control	One or few genes with major effect	Many genes with small but additive effect
Stability	Liable to sudden breakdown by new pathogen races	Not affected by changes in virulence genes of the pathogen
Commonly used but not strict synonyms	Vertical, race-specific, seedling, differential, monogenic	Horizontal, race non-specific, mature plant, adult plant, field, uniform

(After Hayes & Johnston, 1971.)

Although the pathway to durable resistance is very narrow, there have been some outstanding successes. The winter wheat variety Cappelle-Desprez, which occupied more than 80% of the total wheat area in Britain for 10 years and was a widely grown variety for over 20 years, was adequately resistant to both yellow rust and eyespot caused by *Pseudocercosporella herpotrichoides*. It was subsequently used as a source of yellow rust and eyespot resistance in many modern wheat varieties.

A summary of the characteristics of major gene and polygenic resistance is given in Table 5.9.

Use of resistant varieties
In the preceding text, the basis for genetic resistance in plants has been considered. But how do these often complex and sometimes theoretical issues relate to the farmer? How can a farmer best deploy resistant varieties on his farm?

The first and most obvious point is that resistance is available in many existing crop varieties. In the UK, the NIAB annually produces a series of farmers' leaflets for all the major arable crops. Contained in each leaflet is an assessment of the yield, quality, agronomic features and disease resistance of popular, NIAB-recommended varieties. Resistance ratings are most commonly given on a 1–9 scale where the highest rated variety has most resistance. For example in Farmers' Leaflet No. 8, *Recommended Varieties of Cereals* (1989), 1–9 resistance ratings are given for the diseases mildew, yellow rust, brown rust, *Septoria nodorum*, *S. tritici*, eyespot, sharp eyespot, *Fusarium* ear blight and loose smut of 14 varieties of winter wheat. Resistance ratings are given in other leaflets for major diseases of potatoes, oilseed rape, sugar beet, peas and beans, fodder crops, some vegetables and grasses. The farmer thus has a considerable selection of varieties with different resistance ratings to choose from. Other features of varieties, most notably yield and, especially in today's market, quality, will probably influence the farmer's decision more than resistance to diseases. However, farmers are becoming increasingly aware of the effects of severe disease epidemics on both yield and quality of a crop. Faced with resistance ratings to all major diseases, the farmer has to decide which diseases are most important on his farm. He can then, considering other requirements of a variety, select varieties with a high resistance rating to those diseases which he knows, from past experience, are likely to cause him trouble. This is clearly the most direct way that a farmer can deploy resistant varieties to control diseases. In addition there are other, less obvious ways in which genetic resistance in varieties can be used.

Most farmers now use fungicides to control diseases, particularly in cereals. This does not necessarily mean that resistant varieties are redundant. A farmer who selects a variety primarily because of its quality or agronomic features can then use the resistance ratings as a predictive tool for fungicide usage. Varieties with high resistance ratings are less likely to need fungicides than those with lower ratings.

Diversification of varieties

A further use of genetic resistance in varieties is the practice of growing a selection of varieties on a farm in order to reduce disease spread. If the specific resistance genes in each variety selected differ, a pathogen race which possesses specific virulence genes for one variety will not spread to the other varieties. Diversification is used primarily in cereal varieties for reduction of mildew and yellow rust. Varieties are selected according to 'Diversification Schemes' which can be found in the NIAB Farmers' Leaflet No. 8, *Recommended Varieties of Cereals*. The genetic resistance of varieties is deduced by the UKCPVS using the gene-for-gene hypothesis, as previously described.

The advantage of using diversification schemes is a general reduction in disease on a farm with little or no extra cost to the farmer. There are, however, some constraints to diversification. The first and perhaps the biggest constraint concerns the varieties themselves. Farmers may find it difficult to obtain a combination of varieties which meets their yield, quality and agronomic requirements yet is also well diversified. If a farmer does overcome this first problem, there are others which may be encountered. One or more of the varieties that he selects may be only 'Provisionally Recommended' by the NIAB. As such, seed stocks will probably be limited and the local merchant may not be able to supply his requirements. Assuming this hurdle is overcome, a good diversification of varieties on one particular farm may not be a good diversification with varieties on a neighbouring farm. It could be that the same variety is selected for sowing in adjacent fields by two neighbouring farmers.

Variety mixtures

A logical extension of diversification of varieties is to diversify within a single field. Mixtures or blends of cereal varieties are marketed by a few seed companies in the UK and occupy a relatively small but consistent share of the cereal seed market. Experimental work on mixtures indicates that they usually yield at least as well as, and frequently higher than, the mean yields of the component varieties. There are two main reasons for the yield premium. Firstly, as with diversification between fields, a good variety mixture will consist of components which have different major gene resistances to diseases such a mildew and yellow rust. The progress of disease epidemics is hence slowed down. Wolfe (1985) reported that good mixtures of spring barley varieties resulted in up to 80% reduction in powdery mildew compared with mean disease levels of the components grown as pure stands. There is also evidence that infection by non-specialised pathogens such as *Septoria nodorum* is reduced in mixtures of wheat varieties (Jeger, Jones & Griffiths, 1981). Secondly, even in the absence of disease, mixtures frequently contain varieties which have slightly different growth characteristics. For example, leaf angles, plant heights and root profiles may vary between the component varieties. It is likely that such biological diversity will lead to less interplant competition and more effective use of environmental resources such as sunlight and water. Therefore there may be a small yield premium in mixtures, even when diseases are absent. A further claimed advantage in growing mixtures is stability of yield. Although predictions of yield are useful, they are very difficult to make with single varieties. Extreme environmental factors such as frost and drought are the main reason for fluctuating yields of pure stands over years. In variety mixtures, it is unlikely that all the components will be equally affected by environmental extremes. Consequently, the yield of variety mixtures over years may be more consistent and therefore more predictable.

After a consideration of the many advantages of using variety mixtures of cereals on the farm, it is perhaps a little difficult to understand why more farmers do not grow mixtures. As with diversification, perhaps the biggest problem with mixtures is finding the correct combination of varieties. In fact, as far as millers and maltsters are concerned, there is as yet no right combination. All the essential characteristics of grain for breadmaking or malting can be met in an individual variety, but not when varieties are mixed. However, there are a number of commercially available 'blends' which are successfully sold to the feed wheat market. A further, at present academic, concern about variety mixtures regards the development of 'super-races' of pathogens. In principle, exposing pathogen races to an array of genetic resistances could result in the production of isolates with all the necessary virulences to attack the variety mixture and hence any other variety as well. There is as yet little practical evidence to support this theory. The final disadvantage is that of cost. Variety mixtures are approximately 5–7% more expensive than single varieties. Although there is the likelihood of reduced fungicide costs later in the season, perhaps this balance is overlooked at the beginning of the season when varieties are being selected.

In conclusion, cereal variety mixtures offer comparatively cheap, environmentally acceptable disease control. In situations where inputs are low or fungicides cannot be used, e.g. organic farming, variety mixtures should be seriously considered.

Multilines

In practice, multilines are very similar to mixtures. A multiline is a combination of almost genetically identical breeding lines (isogenic) which have all agronomic characters in common, but differ in major gene resistance. Each component of a multiline is bred by a programme of back-crossing a resistance source to a common parent. A multiline variety consists of several individual components in various proportions. The main advantage of multilines over mixtures is uniformity. The 'variety' differs only with regard to genetic resistance, therefore quality and maturity, for example, are consistent within the variety. The primary disadvantage with multilines is the extra effort required by the plant breeder. This fact, together with the problems in officially registering a cereal multiline as a 'variety' in the UK, at present result in little enthusiasm for the production of cereal multilines.

Mechanisms of resistance in plants

The mechanism by which plants resist disease is a fascinating subject and one which has been very widely investigated. There are, however, few if any incontrovertible conclusions as a result of all this activity. Perhaps this is not surprising when one considers how little understanding there is about plant physiology generally. Adding a plant pathogen to the sometimes confused situation results in further complications. There are, however, a few, fairly well substantiated theories about the basis for major gene and polygenic resistance. Resistance mechanisms are divided by most authors into **passive** and **active** (Figure 5.3).

Passive resistance mechanisms are most commonly associated with polygenic resistance. Passive resistance to plant pathogens is a very common phenomenon which exists normally in a healthy plant and is not produced as a result of a stimulus from an invading pathogen. Passive resistance mechanisms may be either structural or chemical. Structural barriers to pathogen entry and colonisation include the waxy cuticle, the relatively tough cellulose-based plant cell walls and the endodermal layer in plant roots. Stomata and lenticels are entry sites for a number of plant pathogenic fungi and bacteria. However, in the case of stomata, there is some evidence that stomatal structure and behaviour can affect infection. Stomata may be closed during periods of optimum germination and growth of the pathogen. In addition to structural defences, there exist naturally in most plants compounds which can deter plant pathogens. Concentrations of nutrients such as amino acids, proteins and carbohydrates in plants may not be high enough to support certain pathogens, particularly biotrophs. The pH of plant tissue may be unfavourable for growth of a pathogen, especially a bacterial pathogen which is likely to have a narrow range of pH tolerance. Natural plant metabolites such as phenolic substances, alkaloids and glycosides have all been implicated as pathogen deterrents. Finally, certain enzymes produced by pathogens, which are essential to separate and break down tissue, may be inactivated by specific compounds in plants. For example, pathogen-produced pectolytic enzymes can be inactivated by certain glycoproteins located in the host cell wall. All the forementioned passive resistance mechanisms primarily serve purposes in the plant other than resisting disease.

Active resistance mechanisms operate specifically in response to pathogen invasion and are most commonly, although not exclusively, associated with major gene resistance. Structural resistance mechanisms involve attempts by the plant to deter infection and colonisation by the production of physical barriers (Figure 5.4). Callose is a common component of thickened cell walls which may be formed in certain pathogen/host interactions, for example

Figure 5.3. Classification of the main resistance mechanisms in plants. (After Dickinson & Lucas, 1982b.)

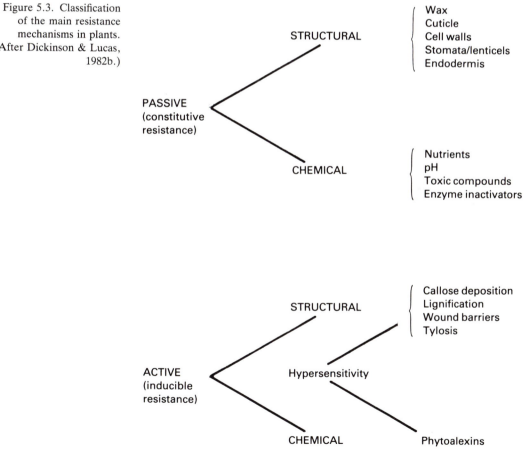

Figure 5.4. Structural
resistance mechanisms.

Lignitubers

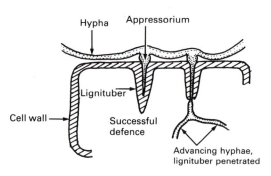

Wound barriers

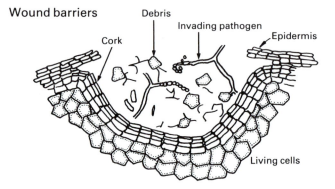

Tylosis

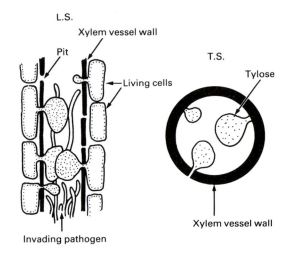

thickened cell walls are produced in resistant cucumber varieties when challenged by *Cladosporium cucumerinum* which causes scab. Lignin may form in response to infection either as a thickened cell wall, or as a specific deposit together with other chemicals such as cellulose and callose around an invading hypha (**lignituber**). Although there is little field resistance to the take-all pathogen *Gaeumannomyces graminis*, lignitubers have been frequently observed surrounding invading fungal hyphae. In the most susceptible reactions, fungi appear to have the ability to grow through these barriers. When plant cells are damaged, whether mechanically or by pathogens, a series of responses is initiated, resulting in the formation of **wound barriers**. Damaged cells die and in the adjacent layers of cells, substances such as suberin, lignin gum and tannin accumulate. A few days later, a cork layer is formed which may restrict further invasion by pathogens. Wound barriers occur in common scab of potato tubers. The disease name is derived from the wound barrier, or scab, which forms in response to attempted invasion by the pathogen. The corky barrier is, in this particular instance, sufficient to stop further invasion of tissue and common scab is only a superficial skin disorder. **Tyloses** are rather more specialised structures produced mainly in response to vascular wilt diseases. Tyloses are balloon-like intrusions into the lumen of a xylem vessel, formed by enlargement of an adjacent living cell and its extrusion through a pit in the vessel wall. This physical blocking of the xylem vessel produced in response to infection by vascular wilt pathogens such as *Verticillium albo-atrum* and forma speciales of *Fusarium oxysporum* may serve to restrict fungal colonisation of the xylem.

Specific anti-pathogenic chemicals may be formed by plants in response to many stimuli (chemical, physical, biological) including pathogen invasion. These are called **phytoalexins**. Such compounds have been isolated from a range of crop plants, including potatoes (rishitin, chlorogenic acid, phytuberin), French beans (kievitone, phaseollin) and peas (pisatin), primarily after plants have been challenged by specific pathogens. The significance of phytoalexins in disease resistance is debatable. The concentration of phytoalexin is generally very low in a plant but may accumulate around an infection site where it could be effective in reducing disease. There is evidence to suggest that in some pathogen/host combinations, phytoalexins are the basis of the major gene resistance. In the case of crown rust of oats caused by *Puccinia coronata*, Mayama *et al.* (1982) suggested that an oat phytoalexin, avenalumin, was found in the highest concentration in race/host combinations with the greatest incompatibility (avirulent combinations).

A common, better understood, resistance mechanism which involves both chemical and structural responses is **hypersensitivity**. A hypersensitive response in plants involves very rapid death of cells penetrated by the pathogen. In addition, cell death is frequently accompanied by other changes aimed at reducing pathogen spread. Adjacent cell walls may lignify and toxic compounds and phytoalexins may also accumulate. For biotrophic fungi, the hypersensitive response is particularly important in that it deprives the pathogen of the living cells required for its nutrition. However, the response has also been observed in attempted invasions of tissue by necrotrophic fungi and bacteria. Hypersensitive reactions are frequently found in cereal varieties which possess major gene resistance to powdery mildew. They are commonly seen as minute chlorotic or necrotic flecks on leaves, with no apparent cause.

The whole subject of resistance mechanisms in plants is very complex, frequently involving many of the structural and biochemical processes briefly described here. When we understand these processes more fully, resistance mechanisms may develop from a largely laboratory-based aspect of plant pathology into a practical proposition for farmers. Perhaps one day farmers will be able to immunise crops against disease by spraying with some sort of 'resistance inducing factor'.

Chemical control

Introduction
At present, many farmers seem to equate control of plant diseases with use of fungicides. This is understandable as, for certain diseases, the use of fungicides is the most effective and indeed often dramatic way of reducing disease. However, fungicides should ideally be used as part of an integrated approach to disease control. In some cases, fungicides should be used as a last resort; when cultural, legislative and biological control together with resistant varieties have been found wanting, fungicides should be applied.

History of chemical control
The first milestone in the history of fungicide discovery was probably the use of lime–sulphur in 1802 by the Royal gardener, Forsyth, for control of mildew in fruit trees. Copper sulphate was used as a seed treatment in 1807 by Prevost for the control of bunt of wheat and lime–sulphur in 1880 by Ward in an attempt to control coffee rust. One of the most famous accounts of copper sulphate

and lime usage concerned the control of vine downy mildew. Vine-growers in Bordeaux, France were having problems with people stealing grapes from parts of the vineyard which bordered a road. In order to deter them, a mixture of copper sulphate and hydrated lime was sprayed over the vines which were most at risk from theft. The chemical mixture gave the grapes an unappetising blue colour and immediately reduced the amount of theft. The Professor of Botany at Bordeaux University at the time, Millardet, observed that vines treated with the mixture did not suffer serious attacks of downy mildew, which was a considerable problem in most vineyards. Realising the significance of the discovery, Millardet went on to demonstrate the effectiveness of 'Bordeaux mixture' in a field experiment in 1885.

The next major step was the introduction of organomercurial seed treatments for control of bunt of wheat by Riehm in 1913. Such treatment, although still used today in many countries, is becoming less popular as a result of legislation.

In 1934, the Du Pont company discovered the first 'organic' fungicides, the dithiocarbamates. Over the subsequent 10 years products such as thiram, zineb and maneb were developed. However, like the earlier fungicides, the dithiocarbamates are 'protectant' or 'surface' fungicides. They do not penetrate plants, cannot cure disease and are subject to weathering. It was not until the late 1960s that fungicides which could enter plants and be translocated, the so called 'systemic' fungicides, were developed. Several chemicals, including the carboxamides, benzimidazoles, pyrimidines and morpholines, appeared over a relatively short period. The era of highly effective fungicides with low phytotoxicity and curative action had begun.

The next important discoveries led to the development, in the mid-1970s, of the phenylamides, dicarboximides, triazoles and fosetyl-aluminium, giving farmers and growers a considerable range of weapons in their armoury to fight disease [for further information on the history of fungicides, see Ainsworth (1981) and Brent (1985)].

Since the mid-1970s, no new fungicide groups have been made commercially available although a number of new compounds have been produced from molecules within existing fungicide groups. This could become significant in view of the fact that some important plant pathogens have developed resistance to fungicides.

Methods of chemical control

There are three general methods of chemical control: **prophylaxis**, the healthy host is protected; **chemotherapy**, the diseased host is cured and **disinfestation**, spores or pathogen propagules are destroyed either on the surface of the host or in the environment surrounding the host. The activity of fungicides may also be divided into three basic categories: **protectant**, **preventative** or **contact** fungicides, used in prophylaxis; **eradicant**, **curative** or **systemic** fungicides, used in chemotherapy and **disinfectants**, used in disinfestation. Any particular fungicide may not fit neatly into one of the above categories. Hence a good fungicide will probably have both protectant and eradicant or systemic activity. For convenience, however, the major groups of agricultural fungicides are often divided into either protectant or systemic.

Protectant fungicides do not have the ability to penetrate the plant cuticle and are therefore not translocated within the plant to any great extent. As such, a protectant is best applied before fungal spores reach the plant surface, as it will have a limited effect on disease already present. Protectants can, however, affect sporulation or viability of spores and a number of protectant fungicides have some eradicant activity, particularly against diseases such as powdery mildew which develop primarily on the surface of the leaf. A good protectant fungicide must have the ability to kill pathogen spores quickly yet not harm the plant. Formulations of protectant fungicides must also stick to plant surfaces and be resistant to weathering. Rain showers, however, can be important in the re-distribution of protectant fungicides from the upper to lower parts of the plant canopy. Although protectant fungicides still play a very important role in the protection of many crops world-wide, the mode of action of the majority of the groups remains unclear. Protectants may affect a wide range of metabolic processes in the fungus.

Systemic fungicides are taken up by the plant, translocated to some extent and protect the plant from attack by pathogenic fungi. Most will also be able to eradicate or cure disease to some degree. A candidate systemic fungicide must be capable of being absorbed and translocated. It must also either kill the fungus (fungicidal) or stop fungal growth (fungistatic) but not damage the plant (non-phytotoxic). The advantages of using systemic fungicides over protectants are considerable. Systemics can be applied in one position and be effective elsewhere, for example in a place that might be difficult for a spray to penetrate, and they can protect new plant growth. They are not as subject to weathering as protectants and can also be effective after infection has occurred. Some systemic fungicides may produce vapours which have fungicide activity. This 'vapour activity' results in an increased zone of protection on foliage and is used to good effect in the control of cereal powdery mildews with compounds such as triadimefon.

The uptake and translocation of systemic fungicides is not well understood in many cases. The penetration of plant cuticle, and entry to the **apoplast** (non-living parts of the plant including cell walls and xylem), and **symplast** (living parts including phloem and protoplast) depend on the physico-chemical properties of the fungicide and the host plant. This relationship may also be critical in the subsequent translocation of the compound. Some systemic fungicides have translaminar activity, i.e. when applied to one surface of a leaf they can control disease on the other surface. Most systemic compounds move in the transpiration stream. If roots of some plants are treated with fungicides such as benzimidazoles and phenylamides, the compounds may be transported over long distances to protect young foliage. Very few fungicides move in the phloem of plants, although fosetyl-aluminium appears to be translocated in both the xylem and the phloem. For further details, see Shephard (1985).

Protectant and systemic fungicides can be further categorised according to mode of action. For each fungicide group, active ingredients together with some examples of proprietary products available in the UK are listed below. The current approved uses of the fungicides in UK agriculture are listed in the *UK Pesticide Guide* (Ivens, 1989) and their mode of action briefly described. For further information on uses and application of specific products, it is important to consult the annual product guides produced by agrochemical manufacturers.

Protectant fungicides

Chlorobenzenes

Active ingredient (product)	Crop	Disease
tecnazene (Fusarex, Hytec, Hortag)	Potatoes (stored ware crops)	dry rot
tecnazene + thiabendazole (Storaid dust, Storite SS, Hytec super)	Potatoes (stored ware crops)	dry rot, silver scurf, gangrene, skin spot
quintozene (Brabant PCNB 20%, Tubergran)	Lettuce (soil applied) Potatoes	*Botrytis*, bottom rot, *Sclerotinia* rot *Rhizoctonia*

Mode of action
Unclear, possibly interferes with some aspect of chitin metabolism.

Phthalonitriles

Active ingredient (product)	Crop	Disease
chlorothalonil (Bravo 500, Repulse)	Cereals	*Septoria* spp., *Rhynchosporium*
	Potatoes	late blight
	Oilseed rape	downy mildew, *Botrytis*
	Field beans	chocolate spot
	Vegetable leaf brassicas	*Alternaria, Botrytis*, downy mildew, damping-off, ring spot
	Field vegetables	range of diseases, especially *Botrytis*

Mode of action
Thought to inhibit thiol (sulphydril) groups in dehydrogenase enzymes.

Dithiocarbamates

Active ingredient (product)	Crop	Disease
mancozeb (Dithane 945, Manzate 200, Penncozeb)	Cereals	*Septoria* spp., leaf stripe, mildew, rust, sooty-mould
	Potatoes	late blight
	Oilseed rape	downy mildew
	Lettuce	downy mildew
maneb (Manzate, Trimangol 80)	Cereals	*Septoria*, yellow rust, brown rust, eyespot, net blotch, *Rhynchosporium*, sooty mould
	Potatoes	late blight
	Oilseed rape	*Alternaria*, downy mildew
	Brassicas	*Alternaria*, downy mildew
	Sugar beet	leaf spot
manganese zinc ethylene-bisdithiocarbamate: available only in mixtures (see ofurace).		
zineb (Hortag)	Potatoes	late blight
	Onions and leeks	downy mildew
	Lettuce	downy mildew

| zineb-ethylenethiuram disulphide complex (Polyram) | Potatoes | late blight |
| | Oilseed rape | downy mildew |

Mode of action
Similar to phthalonitriles.

Organomercury

Active ingredient (product)	Crop	Disease
phenylmercury acetate (Agrosan D, Ceresol, Mist-O-matic mercury)	Wheat (seed)	bunt
	Barley (seed)	covered smut, leaf stripe, net blotch
	Oats (seed)	covered smut, leaf spot, loose smut
	Fodder beet (seed)	blackleg

Mode of action
Unclear, probably mercury ions which are fungitoxic. Mercury is toxic to many biological systems.

Organotin

Active ingredient (product)	Crop	Disease
fentin hydroxide (Du-Ter 50, Farmatin 50)	Potatoes	late blight
	Sugar beet, seed crops	*Ramularia* leaf spot
fentin acetate + maneb (Brestan 60)	Potatoes	late blight

Mode of action
Unclear, may inhibit respiration and attack membrane integrity.

Sulphur

Active ingredient (product)	Crop	Disease
sulphur (Thiovit, Magnetic 6, Kumulus S, Solfa)	Cereals	powdery mildew and foliar feed
	Oilseed rape	powdery mildew and foliar feed
	Sugar beet	powdery mildew
	Peas and beans	powdery mildew and foliar feed

Mode of action
Unclear, reduced sulphur (H_2S) may interfere with protein metabolism, respiration and chelate heavy metals within cells.

Systemic fungicides

MBCs (methyl benzimidazole-2-yl carbamate) or Benzimidazoles

Active ingredient (product)	Crop	Disease
benomyl (Benlate)	Cereals	eyespot, *Rhynchosporium*
	Oilseed rape	*Botrytis*, light leaf spot
	Field beans	chocolate spot
	Peas (seed)	leaf and pod spot
	Vegetable brassicas	light leaf spot, ring spot
	Carrot and parsnip (dip)	black rot, *Botrytis, Sclerotinia* rot
	Onions (seed)	neck rot
	Lettuce	grey mould (*Botrytis*)
carbendazim (Bavistin, Derosal Liquid, Focal Flowable, Stempor DG)	Cereals	eyespot, *Rhynchosporium*
	Oilseed rape	light leaf spot
	Field beans	chocolate spot
	Dwarf and navy beans	*Botrytis*
	Lettuce	big vein
	Onions and leeks	leaf rot (*Botrytis*)
fuberidazole: available only in mixtures (fuberidazole + triadimenol = Baytan seed treatment)		
thiabendazole (Storite Flowable, Tecto)	Oilseed rape (seed)	canker
	Potatoes (seed and stored ware tubers)	dry rot, gangrene, silver scurf, skin spot
thiophanate-methyl (Cercobin, Mildothane)	Winter wheat	*Botrytis*, eyespot, *Fusarium* ear blight, sooty mould
	Barley	eyespot, powdery mildew, *Rhynchosporium*
	Oilseed rape	*Botrytis*, light leaf spot, *Sclerotinia*
	Field beans	chocolate spot
	Brassicas	clubroot

Mode of action

Members of the group have both protectant and eradicant activity. Both benomyl and thiophanate-methyl are converted to carbendazim in plants and it is widely believed that it is this compound which is fungitoxic. The toxicity is due primarily to an interference with the formation of tubulin which is normally assembled into the microtubules (these make up the spindle fibres essential in mitosis). Many fungi have developed resistance to MBC fungicides.

DMI group of ergosterol biosynthesis inhibitors

Active ingredient (product)	Crop	Disease
imazalil (Fungaflor C)	Potatoes (seed)	silver scurf, skin spot, gangrene, dry rot (limited use)
prochloraz (Sportak)	Cereals	eyespot, mildew, *Septoria* spp., *Rhynchosporium*, net blotch, sharp eyespot
	Oilseed rape	*Alternaria*, light leaf spot, *Phoma* stem canker, white leaf spot, *Botrytis* grey mould, *Sclerotinia* stem rot
triforine (Saprol)	Barley	powdery mildew, net blotch
nuarimol (Triminol)	Cereals	powdery mildew, *Rhynchosporium*, net blotch, *Septoria*
flutriafol: available only in mixtures (flutriafol + chlorothalonil = Impact Excel)		
propiconazole (Radar, Tilt)	Cereals	powdery mildew, brown rust, yellow rust, eyespot, *Rhynchosporium*, net blotch, *Septoria*, sooty mould
	Oilseed rape	light leaf spot, *Alternaria*
triadimefon (Bayleton)	Cereals	powdery mildew, *Rhynchosporium*, rust
	Beet	powdery mildew
	Vegetable brassicas	powdery mildew
	Leeks	rust
triadimenol (Bayfidan)	Cereals	powdery mildew, rust, *Rhynchosporium*, snow rot
	Beet	powdery mildew
	Vegetable brassicas	powdery mildew

Mode of action
Many members of the group have both protectant and eradicant activity. The formation of ergosterol, a major sterol in many fungi

which plays an important role in the structure and function of fungal membranes, is inhibited. Fungi-toxic methylated sterols also accumulate (Figure 5.5). Some fungi, particularly formae speciales of *Erysiphe graminis*, which cause cereal powdery mildews, have developed resistance to fungicides in this group.

Morpholine group of ergosterol biosynthesis inhibitors

Active ingredient (product)	Crop	Disease
fenpropimorph (Corbel, Mistral)	Cereals	powdery mildew, yellow rust, brown rust, *Rhynchosporium*
	Field beans	rust
	Leeks	rust
tridemorph (Bardew, Calixin, Ringer)	Cereals	powdery mildew
fenpropidin (Patrol)	Cereals	powdery mildew, yellow rust, brown rust, *Rhynchosporium*, *Septoria nodorum*

Mode of action
Morpholines also inhibit ergosterol biosynthesis in fungi but do so at different points in the synthetic pathway (Figure 5.5). Resistance of powdery mildews to the morpholines has not developed to any extent so far and morpholines can control isolates of *E. graminis* which have become resistant to some of the triazoles.

Figure 5.5. Biosynthetic pathway of ergosterol. ■, point of blockage (site of action) of triazoles (demethylation inhibitors = DMI group); □, points of blockage of morpholines. (After Sisler & Ragsdale, 1984.)

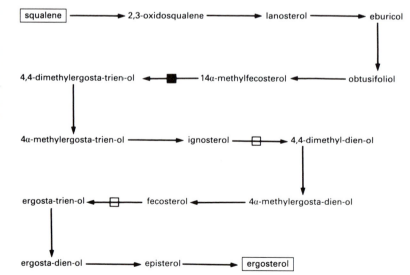

Hydroxypyrimidines

Active ingredient (product)	Crop	Disease
ethirimol	Barley	powdery mildew
(Milstem)	(seed)	

Mode of action

There is evidence that ethirimol inhibits the activity of adenosine deaminase, an essential enzyme involved in nucleic acid synthesis. After the widespread use of ethirimol, resistant strains of *E. graminis* appeared in the population and the use of the product was restricted to spring cereals to preserve its activity. Later, the availability of alternative fungicides, the triazoles, led to the re-introduction of ethirimol for use in winter crops (see resistance to fungicides).

Carboxamides

Active ingredient (product)	Crop	Disease
benodanil	Cereals	yellow rust, brown rust,
(Calirus)		snow rot (winter barley)

carboxin: only available in
mixtures, e.g. carboxin
+ thiabendazole = Cerevax

Mode of action

The site of action of the carboxamides appears to be the mitochondria where they inhibit succinate oxidation, an essential respiratory process.

Organophosphates and phosphonates

Active ingredient (product)	Crop	Disease
fosetyl-aluminium [phosphonate] (Aliette)	Broad beans	downy mildew
pyrazophos [organophosphate] (Missile)	Barley	powdery mildew, net blotch, *Rhynchosporium*
	Vegetable brassicas	powdery mildew
tolclofos-methyl [organophosphate] (Rizolex)	Potatoes (seed)	black scurf, stem canker
	Lettuce	damping-off

Mode of action

Organophosphates are widely used in crop protection as insecticides. As fungicides, their mode of action is unclear. Tolclofos-methyl is a non-systemic contact fungicide with protective and curative action.

Dicarboximides

Active ingredient (product)	Crop	Disease
iprodione (Rovral Flo)	Wheat	glume blotch (*Septoria nodorum*)
	Barley	net blotch
	Oilseed rape	*Alternaria, Botrytis, Sclerotinia* stem rot
	Field beans	chocolate spot
	Peas	*Botrytis*
	Vegetable brassicas	*Alternaria, Botrytis*
	Onions	Leaf rot (*Botrytis*), neck rot
	Lettuce	*Botrytis*
vinclozolin (Ronilan)	Oilseed rape	*Alternaria, Botrytis, Sclerotinia* stem rot, light leaf spot
	Beans	chocolate spot
	Onions	leaf rot (*Botrytis*), white rot
	Lettuce	*Botrytis*

Mode of action

The two members of this group differ somewhat in their physical properties. Neither is truly systemic although both have limited eradicant activity. The precise biochemical mode of action of the dicarboximides is not well understood. There is some evidence for interference with membranes, cell walls and nuclear processes.

Phenylamides

Active ingredient (product)	Crop	Disease
benalaxyl + mancozeb (Galben M)	Potatoes	late blight
metalaxyl + mancozeb (Fubol 75 WP)	Potatoes	late blight
	Oilseed rape	downy mildew
metalaxyl + mancozeb (Fubol 58 WP)	Vegetable brassicas	white blister
	Beans	downy mildew
	Carrot	cavity spot
	Lettuce	downy mildew

metalaxyl + thiabendazole + thiram (Apron Combi 453 FS)	Peas and beans (seed)	*Ascochyta*, downy mildew, damping-off
metalaxyl + thiram (Favour 600 FW)	Lettuce	downy mildew
ofurace + manganese zinc ethylenebisdithio-carbamate (Patafol Plus)	Potatoes	late blight
	Oilseed rape	downy mildew
oxadixyl + mancozeb (Recoil)	Potatoes	late blight

Mode of action
Unclear. Phenylamides may interfere with RNA synthesis. Fungicide-resistant isolates are present at varying frequencies in most *Phytophthora* populations, hence the inclusion of a non-systemic active ingredient, e.g. dithiocarbamate, in all formulations including phenylamides.

Application and formulation
All the active ingredients listed above are approved for application to the aerial parts of the plant, stored plant material or seed. There are also a few recommendations for fungicide application to soil. The formulation of active ingredients required usually differs in each of these situations. Correct formulation and application are critical to the efficacy of the fungicide.

Application to aerial parts of the plant
Fungicides are most frequently applied aerially; generally, plant foliage is sprayed with the fungicide product suspended in water. Most fungicides are not water-soluble and therefore require surface-active agents that enable the powder to be mixed with water to form a suspension (wettable powders). Alternatively, fungicides may be formulated as emulsifiable concentrates, whose active ingredients are dissolved in organic solvents such as xylene. The emulsifier, a surface-active agent which is partly hydrophilic and partly lipophilic, enables the formation of a homogeneous and stable dispersion of small globules of the solvent in water. Other additives in fungicide formulations include wetters, surfactants and spreaders which are used to facilitate contact between the spray and the foliage. Stickers may also be included in an attempt to keep the product on the plant surface for as long as possible.

The farmer has many choices to make in the application of fungicides, including the volume of spray applied to the crop and the type of spray nozzle used. He may want to consider some of the recent developments in application technique such as controlled droplet application (CDA), electrostatic sprayers and ultra-low volume application (less than 50 l/ha). Droplet size is important if fungicides are to be applied effectively with minimum contamination of the environment. Small droplets are produced in ultra-low volume spraying; these cover foliage surfaces more effectively than larger droplets, but are also more prone to drift and evaporate more quickly. At present in the UK the majority of farmers apply a medium volume of spray to crops (220 l/ha) using a standard 'flat fan' nozzle. It is likely that the efficacy of all but the most potent systemic fungicide could be increased by improved application techniques. For further information on application methods see Matthews (1979).

Application to seed

Traditionally, fungicide seed treatments have been used as disinfectants. They have been aimed at killing pathogens such as *Fusarium* spp. and *Pyrenophora graminea* (the cause of leaf stripe) on cereals, which contaminate the surface of the seed. However, recently developed systemic seed treatments such as Baytan (triadimenol + fuberidazole), Ferrax (flutriafol + ethirimol + thiabendazole) and Aliette Extra (captan + fosetyl-aluminium + thiabendazole) will kill pathogens both on the seed surface and inside the seed, and thus give some protection to the seedling from airborne disease attacks for the first few weeks of growth. There is, however, a price to pay for such added protection. Systemic seed treatments are much more expensive than traditional organomercurial seed treatments and they affect the plant physiologically; such effects may be either harmful or beneficial. Baytan-treated wheat plants tend to be more prostrate and greener, which may enable them to survive harsh winter frosts. However, Baytan delays emergence of seedlings, making its use inappropriate with late-sown crops.

Seed treatments may be applied directly as dusts, slurries, or, for a few crops, in seed pellets or in thin films. Seed pelleting is a well established commercial process whose main purpose is to build up small or irregularly shaped seeds into spherical capsules, making them easier to handle and drill. The pelleting material may consist of cellulose powder or chalk together with some binding material. Fungicides can be mixed either throughout the coating material, or in discrete layers. Sugar beet is the major seed species pelleted world-wide and included in the pellet may be phenylmercury acetate, thiram or maneb aimed at reducing soil and seed-borne pathogens such as *Pythium* and *Phoma betae*. Film coating of seed is a new seed treatment technology whereby pesticides are incorporated into thin polymer films which evenly cover the surface of the seed. The advantages of film coating are an evenness of fungicide application, reduced application rates, increased safety both in the application process and for the end-user, and the possibility of the application of high chemical loadings of combined fungicides and insecticides on seeds. A further, and increasing, area where seed is treated with fungicide concerns the production of 'seed' (small tubers) potatoes. Farmers growing their own seed, or buying it from specialist seed producers in either Scotland or Holland, may choose to have it fungicide-treated. The small seed tubers can be treated with a number of compounds including tolclofos-methyl (Rizolex), 2-aminobutane, or thiabendazole (Storite) which are applied either as dusts, fumigants, or slurries. The chemicals can control diseases such as black scurf and stem canker, gangrene, dry rot, silver scurf and skin spot. Seed tubers may become contaminated with fungal pathogens in the field, and diseases may develop in store and be carried over into the subsequent crop.

Application to soil

The application of fungicides to soil is generally aimed at the control of soil-borne pathogens although foliar diseases are also controlled in some cases. For example, primary systemic colonisation by downy mildew in tobacco can be controlled by metalaxyl incorporated into seed beds. There are few other examples of soil-applied fungicides which are selective. Most other methods attempt to sterilise soil and therefore adversely affect beneficial soil-borne saprophytes as well as reducing the pathogen population. A further problem with soil sterilisation is one of scale. The cost of sterilising the soil in a field is usually prohibitive and this is the main reason why soil sterilisation is a rare practice in agriculture. In addition, soil sterilants will not be totally effective, mainly as a result of the impenetrability of the medium they work in, and consequently a reduction in pathogen population, rather than elimination, is the general outcome of the procedure. Soil sterilisation is a much more common practice in horticulture where limited quantities of soil may be effectively sterilised by steam or by general biocides such as methyl bromide, formaldehyde, chloropicrin, metham-sodium or dazomet. Pests and weeds may also be controlled by these treatments. In some areas of Europe, patches of sugar beet fields contaminated by the soil-borne fungus *Polymyxa betae*, the vector of 'rhizomania' disease are fumigated with methyl bromide at a cost of approximately £2000 per hectare (1988).

Application programmes

Decisions regarding the optimum time of application and choice of fungicide may be difficult for the farmer to make. Quite often he will leave them to a specialist crop consultant, ADAS advisor or agrochemical sales representative. Advice is available in a number of forms: 'Counsellor', a computer-based system from ICI; leaflets, for example *Use of Fungicides and Insecticides on Cereals* (Booklet 2257, MAFF, 1986) from ADAS; interactive advisory systems, where observations are sent in and the risks analysed centrally, e.g. EPIPRE (EPIdemic PREdiction and PREvention), a Dutch computer-based pest and disease management system for wheat; or simple 'spray warnings' by post, radio or answerphone. All involve an assessment of the risks of infection and yield loss, coupled with measures which suit the management policy for the particular crop.

Disease risk assessment involves monitoring each field on a regular basis. For cereals, levels of disease are assessed and a decision is made to apply a fungicide only when there is a risk of disease becoming sufficiently severe to cause a yield loss. This can be ascertained following specific guidelines given for most of the important cereal diseases. For some diseases, for example mildew, rust and barley net blotch, crops should be sprayed as soon as a critical level, or threshold level of disease is reached (see sections on individual diseases). Other more weather-dependent diseases such as *Septoria* spp. and *Rhynchosporium* will include weather data in a risk assessment. Eyespot risk assessment may also include cultural factors. Disease risk assessment for potato blight involves warnings of the likelihood of disease development, based on temperature and humidity thresholds in the locality (see late blight of potatoes). In this case fungicides should be applied before disease occurs in the crop. The main advantage of disease risk assessment is that individual fields are routinely monitored for disease, giving a more rational basis for fungicide application. In addition, unnecessary sprays should be avoided and the farmer may save money. The main disadvantage is that too few sprays may be applied. It is possible for the plant to suffer with several diseases simultaneously, each of which does not reach the threshold, but which in combination could reduce yield significantly. A further disadvantage to the farmer concerns planning and management. If he does not know his fungicide requirements until the disease threshold is reached, it may be difficult for him to organise the supply and spraying of the fungicide in the short time available before disease causes damage to the crop. Also, it may be difficult to treat the crop while the weather is wet and favouring the spread of disease.

Routine prophylactic treatment involves routine preventative fungicide applications at set intervals or at certain growth stages of the crop. A decimal code for the growth stages of cereals is given in Table 5.10. There is no monitoring of disease, and fungicides may be applied irrespective of disease levels. The main advantage of a routine prophylactic treatment programme is a simplification in planning and a sure result. Farmers know at approximately what time of year they are going to spray and with what compound. They can also purchase fungicides at best prices. In cereals, satisfactory yield responses can be achieved with simple one, two or three spray programmes, particularly if autumn-sown crops are drilled early in moisture-retentive soils. Prophylactic spraying is particularly useful with varieties which respond well to fungicides. An example of a routine prophylactic programme for winter wheat is given in Table 5.11. In potatoes, a routine prophylactic programme consisting of five or more fungicide applications has been widely adopted to avoid damage caused by late blight in the crop. The main disadvantage of prophylactic spraying is its inflexibility. As it does not take into account disease levels, an unexpected disease epidemic may occur between sprays when the crop is effectively left untreated. This is clearly more of a problem in cereals than potatoes as the time interval between sprays in cereals is usually much greater. Conversely, fungicides may be used when there is little or no risk of a disease outbreak and money may be wasted.

Managed disease control attempts to combine the best aspects of disease risk assessment and routine prophylactic treatment. Hence there are some routine treatments to be given at critical points in the crop development and also some decisions to make based on levels of disease in the crop and the prevailing weather conditions. Managed disease control has been applied primarily to cereals in the UK (MAFF, 1986) although the principles apply to all crops. Farmers are generally encouraged by ADAS to adopt managed disease control as it should be the most cost-effective. The ADAS publishes managed disease control systems as a series of flow diagrams in which decisions are made at key growth stages. Cultural

Table 5.10. *Decimal code for the growth stages of cereals[a]*

Germination
00 Dry seed
03 Imbibition complete
05 Radicle emerged from
   caryopsis
07 Coleoptile emerged from
   caryopsis
09 Leaf at coleoptile tip

Seedling growth
10 First leaf through coleoptile
11 First leaf unfolded
12 2 leaves unfolded
13 3 leaves unfolded
14 4 leaves unfolded
15 5 leaves unfolded
16 6 leaves unfolded
17 7 leaves unfolded
18 8 leaves unfolded
19 9 or more leaves unfolded

Tillering
20 Main shoot only
21 Main shoot and 1 tiller
22 Main shoot and 2 tillers
23 Main shoot and 3 tillers
24 Main shoot and 4 tillers
25 Main shoot and 5 tillers
26 Main shoot and 6 tillers
27 Main shoot and 7 tillers
28 Main shoot and 8 tillers
29 Main shoot and 9 or
   more tillers

Stem elongation
30 Ear at 1 cm
31 1st node detectable
32 2nd node detectable
33 3rd node detectable
34 4th node detectable
35 5th node detectable
36 6th node detectable
37 Flag leaf just visible
39 Flag leaf ligule/collar
   just visible

Booting
41 Flag leaf sheath extending
43 Boots just visibly swollen
45 Boots swollen
47 Flag leaf sheath opening
49 First awns visible

Inflorescence
51 First spikelet of
   inflorescence just visible
52 ¼ of inflorescence emerged
55 ½ of inflorescence emerged
57 ¾ of inflorescence emerged
59 inflorescence completed

Anthesis
60
61 } Beginning of anthesis
64
65 } Anthesis half-way
68
69 } Anthesis complete

Milk development
71 Caryopsis watery ripe
73 Early milk
75 Medium milk
77 Late milk

Dough development
83 Early dough
85 Soft dough
87 Hard dough

Ripening
91 Caryopsis hard (difficult to
   divide by thumb-nail)
92 Caryopsis hard (can no longer
   be dented by thumb-nail)
93 Caryopsis loosening in daytime

[a]2-digit code.

(From Tottman & Broad, 1987.)

Table 5.11. *Routine prophylactic programme for winter wheat*

(i)	Seed Use systemic seed treatment to kill seed-borne pathogens, e.g. *Fusarium* spp., *Septoria nodorum*, loose smut, and protect young plant from early infections of mildew, rust and *Septoria tritici*
(ii)	Late October–early November Spray to control aphids in crops at risk to barley yellow dwarf virus
(iii)	First node detectable (GS 31) Spray aimed to control eyespot, but also other diseases, especially mildew and *Septoria* where necessary
(iv)	Flag leaf emergence (GS 39) Spray aimed to control all leaf diseases: mildew, *Septoria* spp., yellow and brown rust
(v)	Ear emergence (GS 59) Spray aimed to control all leaf and ear diseases: mildew, *Septoria* spp., yellow and brown rust, sooty mould, *Fusarium* ear blight

(After MAFF, 1986.)

factors such as sowing date, soil type and previous cropping, varietal resistance, level of disease, previous fungicide treatment and weather are all used at various stages in making the decision to apply a fungicide. An example for winter barley is given in Table 5.12.

As an aid to farmers attempting to practise managed disease control, a number of computerised schemes have been developed. Such schemes produced much interest in the UK when they were first introduced during the early 1980s. However, primarily as a result of financial constraints on both the agrochemical companies who developed some of these systems and the farmers who might consider purchasing them, interest has waned somewhat recently. Computerised schemes are based on the principles used by ADAS in their disease control schemes. The computer puts the expert knowledge (yield loss formulae, epidemiological models, etc.) into the hands of the farmer to apply to his own field and provides guidance on how to assess disease. The systems also help in assessment of risk/benefit to enable the farmer to decide his own policy on fungicide application. Schemes used at present in the UK include 'Counsellor' from ICI and the Dutch EPIPRE. The criteria for decision making for all the above systems are updated regularly as data accumulate and as knowledge of the biology, effects and control of plant pathogens increases (Royle, 1985).

The farmer also has to decide which **fungicide products** he will use. For many diseases, there is a bewildering choice of products all

Table 5.12. *Section of managed disease control for winter barley*

growth stage (GS) 31: First node detectable

The aim is to control eyespot[a] and
leaf diseases.
Fungicides applied at this stage consistently give good yield
responses.

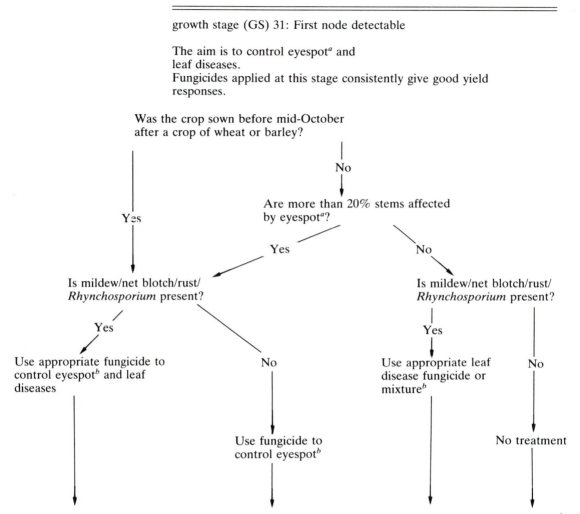

Monitor crop until GS 39; if disease is obvious on any of the top 3 leaves use appropriate fungicide[b]

Next decision GS 39–55

[a]Brown eyespot lesions extend through the two outer leaf sheaths and
mark the sheath beneath.
[b]Consider risk of resistance.
(After MAFF, 1986.)

of which claim to be effective against them. There are several
factors which may help a farmer with his choice. Firstly, a farmer
must ensure that the product is approved for the crop and disease.

In the UK, manufacturers are legally obliged to recommend only products which are approved under the Control of Pesticides Regulations in product manuals and on product labels. Further information regarding spectrum of diseases controlled, dose rates, compatibility, application, harvest interval and toxicity may also influence a farmer's decision. The product manual may also contain some information on the efficacy of the product. Fungicides may be described as, for example, 'suppressing' certain diseases instead of giving 'excellent control'. Perhaps the most reliable source of information regarding fungicide efficacy is the farmer himself. He may have to balance his past experience of fungicide usage with other advice given. Effectiveness of fungicides may change if fungicide-resistant strains build up in the pathogen population so that past performance may not necessarily be a good guide to future effectiveness. The final decision may be based on cost of the fungicide compared with potential benefits of its use.

Costs and benefits

At present almost all farmers in the UK use fungicides. Expenditure on fungicides was £97 million in 1985. Of this, £78 million was spent on cereal fungicides, £2 million on oilseed rape, £5 million on potatoes, £5 million on sugar beet and £7 million on other crops. The benefits in yield and quality in a range of crops following fungicide usage have been demonstrated. Table 5.13 gives an example of yield benefits. Cook & King (1984) proposed that the value of the increase in yield resulting from the application of fungicides to cereals in the UK during 1982 was approximately £118 million. A net benefit of £58 million was calculated after a deduction of the £60 million cost for the fungicides.

As well as reducing disease, there can be other effects from fungicide usage; for example, yields of certain crops, most notably cereals, can increase as a result of fungicide application in the absence of obvious disease. In these cases fungicides may be delaying senescence, stimulating or modifying growth, reducing the populations of 'harmless' surface microflora or of undetectable diseases. Their widespread use suggests that fungicide application is perceived to be financially worthwhile. The most obvious direct cost to the farmer is the price of the fungicide. Generally the older protectant fungicides are considerably cheaper, but produce smaller benefits, than the systemic fungicides. Protectant fungicides may, however, require more frequent use and thereby cost more to apply. Within each group of fungicides, prices differ substantially, as does performance. Tables 5.14 and 5.15 are attempts to produce

Table 5.13. *Yield responses from fungicide treatment of cereal, oilseed rape and perennial ryegrass trials in England and Wales (1981–4)*

	Yield (t/ha)			
Crop	Untreated	Treated[a]	Response[b]	Response (%)
Winter wheat	7.39	8.50	1.11	15
Spring wheat	5.54	6.10	0.56	10
Winter barley	5.94	6.88	0.94	16
Spring barley	5.30	5.95	0.65	12
Winter oilseed rape	3.44	3.83	0.39	11
Perennial ryegrass	15.55	16.10	0.55	4

[a]Fungicide treatment programmes were designed to keep plots free from disease (intensive).
[b]All responses were statistically significant.

(After Priestley, Parry & Knight, 1985.)

Table 5.14. *Cost–benefit analysis of fungicides in wheat*

		Percentage increase in yield required to meet cost[a]	
Fungicide (active ingredient)	Approximate 1988 price (£/ha)	Breadmaking	Feed
Baytan seed treatment (fuberidazole + triadimenol)	12 (170 kg/ha seed rate)	1.5	1.6
Calixin (tridemorph)	11	2.3	2.6
Corbel (fenpropimorph)	20	3.5	3.8
Delsene M (carbendazim + maneb)	10	2.2	2.5
Patrol (fenpropidin)	20	3.5	3.8
Sportak (prochloraz)	20	3.5	3.8
Tilt turbo (propiconazole + tridemorph)	21	3.6	4.0
Typical 3 spray programme (Sportak, Tilt Turbo, Delsene M) with Baytan seed treatment	86	10.6	11.8

[a]Assumes
   (i) application cost £8/ha for foliar sprays (based on mean of contract and average farmers spraying costs quoted in Nix, 1988);
  (ii) yield of wheat 6.7 t/ha (average yield of wheat in UK between 1982 and 1986 from *Agricultural Statistics, United Kingdom, 1988*);
 (iii) price of breadmaking wheat £121/t, feed wheat £109/t (see Table 5.16).

Table 5.15. *Cost/benefit analysis of fungicides in potatoes*

Fungicide (active ingredient)	Approximate 1988 price (£/ha)	Percentage increase in yield required to meet cost[a]
Rizolex seed treatment (tolclofos-methyl)	26 (2.8 t/ha seed rate	1.0
Bravo 500 (chlorothalonil)	14	0.9
Brestan 60 (fentin acetate + maneb)	9	0.7
Dithane 945 (mancozeb)	5	0.5
Fubol 75 WP (mancozeb + metalaxyl)	18	1.0
Recoil (mancozeb + oxadixyl)	18	1.0
Typical 6 spray programme (Fubol × 3, Dithane × 2, Brestan) with Rizolex seed treatment	99	5.8

[a]Assumes:
 (i) application cost £8/ha for foliar sprays (based on mean of contract and average farmers spraying costs quoted in Nix, 1988);
 (ii) yield of potatoes 36.3 t/ha (average yield of maincrop and 2nd early potatoes in UK between 1982 and 1986 from *Agricultural Statistics, United Kingdom, 1988*);
 (iii) price of potatoes £70/t (average price of potatoes 1985–7 from *Annual Review of Agriculture 1988*, HMSO).

cost–benefit analyses for a range of systemic and protectant fungicides used in wheat and potatoes. The application costs for cereals do not take into account the cost of tramlines, which are often made to facilitate pesticide application and reduce yield by about 3%. Given that, on average a three spray fungicide programme increases yield of winter wheat by 15% (Table 5.13), a typical three spray programme should on average be profitable, particularly for breadmaking wheat. The value of potatoes per hectare is on average approximately three times that of wheat. Hence much smaller yield increases are required to meet the cost of fungicide applications (Table 5.15). The application costs do not include the cost of any damage to the crop by the machinery. Even an intensive six-spray programme would be likely to pay for itself in most years, particularly when the potential damage to the crop by blight is considered. Attacks of blight starting early in the season and reaching 75% foliage affected by the end of July will reduce yields by as much as 50%. In addition, should only a few tubers become affected by blight, crops may be totally unmarketable. Bearing in mind the potential benefits of fungicide application, and the stability of fungicide prices since 1985 (Table 5.16), it is perhaps surprising that the increase in fungicide usage in the early part of the 1980s has not continued in the latter part of the decade. The main reason for this is that while fixed costs such as machinery and labour have steadily increased the value of crops has fluctuated (Table 5.16). Hence farmers are under pressure to reduce all variable costs, including fungicide inputs.

The relatively stable price of fungicides over the last few years does not reflect the rapidly rising costs to agrochemical manufacturers of producing new more effective active ingredients. It may take 6–8 years to obtain product registration and the patent life of a new active ingredient may be only 8–10 years. The costs of the research needed to discover and develop a new fungicide have been quoted by Shephard (1987) as $30 million. Clearly, new fungicides must be aimed at a world market. They are consequently often broad-spectrum products developed for use on major world food crops. Broad-spectrum products eliminate non-target organisms such as natural saprophytic antagonists. This can lead to an imbalance in the ecology of the leaf or soil which results in an 'open door' for other pathogens to enter.

Under certain conditions, some fungicides can be phytotoxic. Phytotoxicity is a feature which is screened out during the development of a fungicide, but if mistakes are made with regard to timing and rate of application, plants may be damaged. Additionally, fungicides must not be applied during or immediately prior to extremes of temperature. Even when damage is not obvious, in certain 'soft' crops such as lettuces, pesticides may reduce yield significantly. In some instances, where pesticide programmes have been replaced by other cultural or biological methods of control, increases in yield and quality have been observed.

Safety to operators, consumers and the environment has always been very important in good agricultural practice. The Food and Environment Protection Act 1985, together with subsequent legislation, states that all spray contractors and farm spray operators born after 31 December 1964 must possess a certificate of competence in the safe use of pesticides by 1 January 1989. To obtain a basic certificate of competence, tests must be passed on the safe handling, mixing and storage of pesticides and their potential environmental hazards. Candidates may then progress to tests involving specialist apparatus, e.g. ground crop sprayer–boom type hydraulic nozzle. Such a comprehensive training scheme will allay

Table 5.16. *Relationship between fungicide sales, costs, value of crop (wheat) and fixed costs in UK agriculture in 1980s*

Year	Fungicide sales[1] (Agriculture + Horticulture) (£m)	Fungicide costs[2] (£/ha)		Price of wheat[3] (£/t)		Fixed costs[4] (£/ha)
		Benomyl	Triadimefon	Breadmaking	Feed	
1982	59.3	6.00	13.40	117.5	107.7	360
1983	75.9	5.50	12.00	120.3	113.9	390
1984	90.2	5.20	12.75	135.9	121.1	410
1985	98.1	5.20	14.20	120.4	107.2	420
1986	94.5	5.50	14.80	130.4	106.5	445
1987	97.7	5.80	15.60	121.4	109.0	490

Sources:
1. *British Agrochemicals Association Annual Report 1987/8*, Peterborough.
2. Nix (annual), *Farm Management Pocketbook*, Wye College, University of London.
3. Home Grown Cereals Authority Statistics (1987), Hamlyn House, Highgate Hill, London.
4. Nix (annual), *Farm Management Pocketbook*, Wye College, fixed costs of mainly cereal farm, 100–200 ha.

some of the general public's concern about the misuse of pesticides. Complementary to this legislation is the Control of Pesticides Regulations 1986, which makes it an offence to sell, supply, store, advertise or use unapproved pesticides. To obtain approval, agrochemical companies have to supply exhaustive data to the Ministry of Agriculture on such areas as the performance of the fungicide; residues in food and the environment; rates of break-down; the nature, toxicity and behaviour of metabolites; toxicity to birds, insects, fish and mammals; effects on soil micro-organisms; accumulation in food chains and entry into ground water.

Resistance to fungicides

When a fungus which is sensitive to a particular fungicide becomes less sensitive due to selection or mutation following exposure to the compound, this is called fungicide resistance (after Dekker, 1984). Soon after the introduction of systemic fungicides in the early 1970s, failures to control certain diseases were reported. Table 5.17 lists reported occurrences of fungicide resistance with a wide variety of pathogens, some of which will be discussed in Section II. In contrast to the rapid appearance of resistance to some of the modern systemic fungicides, there are only one or two reported cases of resistance to traditional copper fungicides and dithiocarbamates (protectants) after many decades' usage. The ability of the fungus to develop resistance seems to depend partly on whether the fungicide acts at a single or multiple biochemical sites. Dekker (1984) proposed that fungi may become resistant to fungicides by the following mechanisms:

1. a change at the site of inhibitor action which results in a decreased affinity to the fungicide;
2. decreased uptake or decreased accumulation of the fungicide in the fungus;
3. detoxification of the fungicide before the site of action has been reached or lack of conversion of a compound into the fungitoxic principle;
4. compensation for the inhibitory effect, e.g. by an increased production of an inhibited enzyme;
5. circumvention of the blocked site by the operation of an alternative pathway.

Depending upon the fungus, resistant isolates may arise relatively easily by the mutation of a single gene. Hence fungicides acting at a single site, most commonly the systemic fungicides, are more prone to resistance problems. It is much more unlikely that two or more mutations will occur simultaneously in order for pathogens to

Table 5.17. *Reported cases of fungicide resistance in plant pathogens*

Fungicide group	Pathogen	Disease
Systemic		
'MBC' or benzimidazoles	*Pseudocercosporella herpotrichoides*	eyespot of cereals
	Fusarium nivale	Foot rot of wheat
	Fusarium culmorum	Foot rot of wheat
	Botrytis cinerea	Grey mould of wheat, grey mould of some other horticultural crops
	Botrytis cinerea/ Botrytis fabae	Chocolate spot of beans
	Septoria tritici	Leaf spot of wheat
	Helminthosporium solani	Silver scurf of potatoes
	Pyrenopeziza brassicae	Light leaf spot of oilseed rape
	Cercospora beticola	Leaf spot of beet
Triazoles	*Erysiphe graminis* f.sp. *hordei*	Powdery mildew of barley
	Erysiphe graminis f.sp. *tritici*	Powdery mildew of wheat
	Pyrenophora teres	Net blotch of barley
Hydroxypyrimidine	*Erysiphe graminis* f.sp. *hordei*	Powdery mildew of barley
Carboxamides	*Ustilago nuda*	Loose smut of barley
	Ustilago maydis	Blister smut of maize
Dicarboximides	*Botrytis cinerea/ Botrytis fabae*	Chocolate spot of beans
	Botrytis cinerea	Grey mould of number of horticultural crops
Phenylamides	*Phytophthora infestans*	Late blight of potatoes
	Bremia lactucae	Downy mildew of lettuce
	Peronospora parasitica	Downy mildew of brassicas
Protectant		
Organomercury	*Pyrenophora avenae*	Leaf spot of oats
	Pyrenophora graminea	Leaf stripe of oats
Organotin	*Cercospora beticola*	Leaf spot of beet

become resistant to multi-site fungicides (mainly protectants). Single-gene mutations may be induced in the laboratory by ultra-violet radiation or mutagenic chemicals. However, in the field, fungicide-resistant cells may appear spontaneously by natural mutation, at a rate believed to be between 1 in 10^{-4} and 1 in 10^{-9}. Hence, even in populations of certain pathogens which have never been exposed to a specific fungicide, there may be a small proportion resistant to it. When a fungicide is applied to a crop it will readily kill or inhibit all but the resistant strains of the pathogen. This now allows the resistant strains to multiply. However, in a number of cases, the resistant strains may prove to be less 'fit' than the sensitive strains. They may be less aggressive and slower growing. Hence, by the time the effects of the fungicide wear off (a matter of a few weeks), the more aggressive, sensitive strains may be re-established as the dominant partners in the pathogen population. This suggestion of a reduction in fitness in resistant isolates varies with the fungus, but it is unusual for resistant strains to dominate the pathogen population after a few applications of fungicide within one season. Problems frequently occur after a few years of consecutive applications with the same product, perhaps with short intervals between sprays in one season. During this constant bombardment with fungicide, the sensitive isolates do not have time to re-establish themselves as dominant in the pathogen population. In the constant presence of the fungicide, the resistant strains multiply faster than the sensitive and a critical stage will be reached in the balance of resistant/sensitive isolates, where control is no longer effective in the field (Figure 5.6). In the case of isolates of *Pseudocercosporella herpotrichoides* (the cause of eyespot of cereals) resistant to MBC fungicides, 30% resistant isolates in a field results in ineffective control (King & Griffin, 1985).

Where a fungus develops resistance to a particular biochemical mode of action, similar resistance is usually exhibited towards other fungicides with the same mode of action. This phenomenon is known as positively correlated cross-resistance. For example, isolates of *Phytophthora infestans* resistant to the phenylamide metalaxyl are also resistant to other phenylamides, benalaxyl, ofurace and oxadixyl. Sometimes, resistance to one fungicide is correlated with sensitivity to a second fungicide and vice versa. This is known as negatively correlated cross-resistance. For example, powdery mildew of cereals, particularly barley, has recently shown an increase in resistance to the triazoles. This has been accompanied by an increase in sensitivity to another mildewicide, ethirimol. Resistance to ethirimol was detected in the early 1970s after only a

Figure 5.6. Schematic illustration of build-up of fungicide resistance in pathogens. S, isolates of a pathogenic fungus *sensitive* to the fungicide; R, isolates of a pathogenic fungus *resistant* to the fungicide.

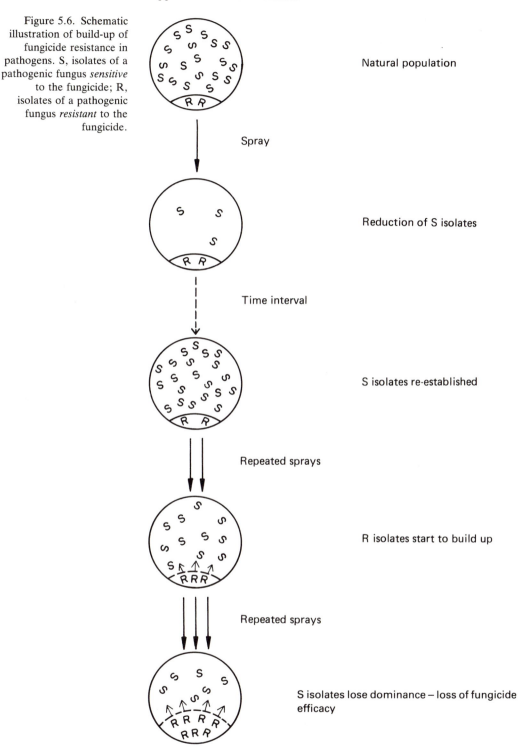

Natural population

Spray

Reduction of S isolates

Time interval

S isolates re-established

Repeated sprays

R isolates start to build up

Repeated sprays

S isolates lose dominance – loss of fungicide efficacy

few years' widespread use in the control of barley powdery mildew and its usage was restricted to preserve its efficacy. Subsequently, ethirimol was largely replaced by the broader spectrum triazoles, but it is now being used in partnership with triazoles to combat resistance to the latter.

It is in the interests of the whole agricultural industry to 'manage' resistant isolates of important plant pathogens effectively and at the same time preserve the available fungicides. To this end, ADAS (MAFF, 1986) have produced a number of strategies for reducing the speed of development of fungicide resistance in cereal pathogens.

1. *Reduce disease severity by non-chemical means*: use other means of controlling diseases such as growing resistant varieties and cultural control.

2. *Diversify fungicide usage*: plan a fungicide programme which does not involve consecutive applications of products with the same mode of action. Also ensure that systemic seed treatments are not followed up with foliar sprays of the same mode of action. If possible, include some multi-site fungicides.

3. *Use appropriate mixtures of active ingredients*: the risk of the development of resistant strains is reduced if active ingredients with two modes of action are mixed. The pathogen then has to mutate to overcome two modes of action simultaneously. Farmers can mix fungicides themselves using compatibility information in product manuals. Many chemical companies now produce ready-formulated mixtures. As well as increasing the potential spectrum of activity, fungicide mixtures also help to safeguard the manufacturer's active ingredients. Mixtures may be of site-specific chemicals with different modes of action, or involve multi-site components. Examples used on cereals include Sportak Alpha (prochloraz and carbendazim), Impact Excel (flutriafol and chlorothalonil), Tilt Turbo (propiconazole and tridemorph), and Delsene M (carbendazim + maneb).

4. *Avoid uneconomic spraying*: fungicides should not be used when there is little disease and only a small chance of net return. Very early and late applications should be avoided if possible. In general, routine prophylactic programmes involving more than three sprays should also be avoided. Exposure of low levels of the pathogen to fungicides is not only wasteful of money but also gives any resistant isolates a chance to multiply. Fungicides should be used sparingly as part of an integrated control strategy.

Biological control

The concept of biological control is difficult to define. In its broadest sense, anything other than chemical control of plant diseases could be classified as biological control. However, for convenience, disease control strategies by cultural methods and resistant varieties have been discussed separately. According to some of the current leaders in the field of biological control, Cook and Baker (1983a), 'biological control is the reduction of the amount of inoculum or disease-producing activity of a pathogen accomplished by or through one or more organisms other than man'. Biological control frequently involves the exploitation of organisms (usually called **antagonists**) in the environment to decrease the capacity of the pathogen to cause disease. The multitude of methods used in biological control can be broadly divided into two groups. Firstly, antagonists can be directly introduced onto or into plant tissue. Secondly, cropping conditions or other factors can be modified in ways known to promote the activities of naturally occurring antagonists. Direct introduction of antagonists is less common. However, inducing resistance in the host by inoculation with non-pathogenic or avirulent strains of a pathogen has been demonstrated in a number of cases, notably in the use of attenuated strains of viruses to control virulent strains. The tristeza virus of citrus crops is controlled in this way, and tomato can be protected from damage by tobacco mosaic virus by prior inoculation with attenuated virus strains. One of the few registered biological control agents which is currently available commercially in the USA and elsewhere involves an avirulent strain of *Agrobacterium radiobacter* (K84) which protects plants against the crown gall pathogen *Agrobacterium tumifaciens* by pre-treatment of wounds with the antagonist. Strain K84 produces a special type of antibiotic or bacteriocin (a high molecular weight protein) that affects only closely related organisms.

Many more cases of biological control involve antagonists in the environment surrounding the host. The mechanisms by which this type of biological control operates may be complex, particularly when aimed at soil-borne plant pathogens which are part of a diverse soil ecology. Antibiosis, competition and hyperparasitism or predation have been recognised by Baker (1985) as the main ways by which biological control operates outside the host. Antibiosis includes any metabolic product(s) of one organism which directly inhibits or destroys another. There is some debate, however, as to whether antibiotics produced under natural as opposed to laboratory conditions can exert a significant effect upon pathogens in the field.

Competition for substrates is particularly vigorous in soil. Biological control, based on the principle of competition, involves attempts to deprive a pathogen of essential nutrients by encouraging the growth of nutrient-consuming antagonists. This may be achieved by rapid incorporation of large quantities of food material into the soil. For example, the cause of bean root rot, *Fusarium solani* f.sp. *phaseoli* may be controlled biologically by incorporation of cellulose or barley straw into soil contaminated with the resting spores (chlamydospores) of the pathogen. Nitrogen, essential for germination and penetration by the pathogen, is made unavailable; it is rapidly utilised by the large population of saprophytes encouraged by the added substrate. The basis of 'green manuring', perhaps one of the best examples of biological control, is competition. Green manuring is simply the incorporation of succulent green tissues, for example grass cuttings or young soybean plants, into soil. It may be useful in the control of common scab (*Streptomyces scabies*) of potatoes, but no specific organisms have been identified as the primary competitors. Competition by a specific antagonist is the basis for one of the few methods of biological control to be practised commercially in the UK. Pine stumps are liable to infection by the fungus *Heterobasidion annosum*, which may then use the stumps as a food/inoculum base for attacking adjacent trees. Chemical protection of the cut surfaces of stumps may be successful at first, but with time becomes ineffective as new surfaces are exposed. Effective and persistent control is obtained by treating cut surfaces soon after exposure with spores of *Peniophora gigantea*, which is not an active pathogen of pine trees. The fungus colonises the outer tissues more rapidly and thus excludes *H. annosum*. The antagonist *P. gigantea* is now grown commercially and marketed in liquid form which is applied to the cut surfaces of the stumps. The treatment is used routinely on pines in the UK and over 62 000 hectares were inoculated with *P. gigantea* in 1973. A further antagonist to be used commercially in the UK in a similar context is *Trichoderma viride*, sold as 'Binab T'. It is approved for control of silver leaf (caused by *Stereum purpureum*) in fruit trees and ornamental trees. Binab T is applied as a paste onto pruning wounds to protect against silver leaf. *Trichoderma* probably acts in a similar way to *P. gigantea*; it rapidly colonises and protects outer woody tissues.

The mechanism of a further biological control phenomenon known as suppressive soils may also be based on competition. Suppressive soils are soils in which the pathogen may persist, but either causes little or no damage, or causes disease for a short time and then declines. The phenomenon of 'take-all decline' (see section on take-all of cereals, p. 160) occurs as a result of soil suppression. Many soils also appear to be suppressive to *Fusarium*, *Phytophthora* or *Pythium* spp. As such suppression can be destroyed by moist heat treatment, and can be transferred to non-suppressive soils simply by mixing suppressive and non-suppressive soils, most workers agree that the suppression occurs as a result of the build-up of saprophytic competitors or antagonists in the soil. *Pseudomonas fluorescens* has been implicated as the predominant antagonist in soil suppressive to take-all; *Pseudomonas* spp. may be involved in *Fusarium*-suppressive soils, and *Trichoderma viride* in soil suppressive to *Rhizoctonia solani*. In a somewhat different context, *Phialophora graminicola* is thought to be responsible for the delay in occurrence of take-all in cereals following grass crops in Britain. However, it is unlikely that any single organism is solely responsible for suppression and this is probably one of the reasons why attempts to induce suppressiveness in field soils by introduction of antagonists have been largely unsuccessful. In addition, more needs to be known about the reasons why antagonists build up in particular soils.

Hyperparasitism has been implicated as a biological control mechanism based exterior to the host. Most attempts have been directed at destroying sclerotial and mycelial inocula and perhaps the best documented examples of hyperparasitism involves the fungus *Rhizoctonia solani*. Damping-off in radish seedlings was observed to decrease in a system of regular weekly plantings of radish seeds. This was apparently attributable to an increase in population of mycoparasitic *Trichoderma* spp. in the soil (Henis, Ghaffar & Baker, 1978).

There are good reasons for adopting biological control practices in agriculture, mainly because of the recent problems with existing methods of control, particularly fungicides. Soil-borne plant pathogens are extremely difficult to control with fungicides and biological control, where available, may be the only alternative method. Even when effective fungicides are available, biological control may offer some advantages. Firstly, there is an increasing problem with resistance to fungicides (see chemical control, p. 140). Biological control methods may be used together with existing fungicides in integrated control strategies to reduce the risk of resistance building up, or if highly effective, biological control methods may replace fungicide sprays completely. A second important reason for adopting biological control concerns the increasing awareness of safety and toxicity problems of fungicides.

Most fungicides have only a temporary effect and applications need to be repeated. Spray operators are therefore exposed to repeat doses of fungicide, and the chances of residues building up in the environment and the produce are also increased. Biological control usually takes longer to reduce disease, but also usually lasts longer. There should be no problems regarding safety or toxicity as most biological control agents are harmless naturally occurring organisms.

There is a view that a revolution in agricultural practices needs to occur if agriculture is to remain viable in many areas of the world. The problems of soil erosion, increasing energy costs, pollution, residues and resistance to fungicides are some of the areas of immediate concern which have arisen at least in part out of the 'short-term' view of agriculture. Financial pressures often result in farmers of necessity having to concentrate on obtaining a profit from a particular enterprise as quickly as possible. An alternative approach is to adopt a 'long-term' view with less emphasis on immediate profits and more consideration for the environmental impact of farming. Biological control fits well into a long-term, less intensive approach to farming, yet despite all its advantages is not practised widely. There are many recognised and documented examples of antagonists which have been studied in the context of biological control (Table 5.18). About 40 genera of Antagonistae vitae with potential for immediate application as antagonists are catalogued by Cook & Baker (1983b). However, very few of these antagonists are used commercially as biological control agents. In the UK there are probably only two examples, which have already been mentioned, namely *Peniophora gigantea* and *Trichoderma viride*. Neither of these is used in agriculture *sensu stricto*. There are a number of reasons for the scarcity of biological control methods. Firstly, soil, where biological control agents are frequently introduced, is an extremely complex environment where interactions are difficult to predict. Hence many very promising antagonists in the laboratory and glasshouse give inconsistent results in the field. Secondly, biological control is slower to take effect than fungicidal control. There are also problems regarding the production and storage of antagonists as living organisms, in sufficient quantitites for large-scale field use. Large agrochemical companies are, however, at present showing interest in the commercial development of biological control agents, on the basis that many small markets may be as profitable as (and more stable than) single large ones.

The problems regarding the patenting of biological control agents are currently being addressed; living organisms cannot themselves

Table 5.18. *Examples of antagonists studied in the biological control of plant pathogens*

Mechanism	Plant	Plant pathogen	Antagonist
Antibiotic Competition/ antibiosis	Many	Agrobacterium tumefaciens	Avirulent Agrobacterium spp.
	Corn	Fusarium roseum 'Graminearum'	Chaetomium globosum
	Pine	Heterobasidion annosum	Peniophora gigantea
	Various	various fungi	Trichoderma spp.
	Various	various fungi	Bacillus subtilis
	Carnation	Fusarium oxysporum f.sp. dianthi	Alcaligenes spp.
	Cotton, Wheat	Pythium, Gaeumannomyces graminis var. tritici, Pseudomonas tolaasii, Fusarium oxysporum f.sp. lini	Pseudomonas spp.
	Apple	Erwinia amylovora	Erwinia herbicola
	Tobacco	Pseudomonas solancearum	Avirulent strain of P. solanacearum
Competition for attachment sites	Many	various fungi	Gliocladium spp.
	Many	Agrobacterium tumefaciens	Avirulent Agrobacterium spp.
Cross protection	Sweet potato	Fusarium oxysporum f.sp. batatas	Non-pathogenic F. oxysporum
	Cucurbits	Fusarium solani f.sp. cucurbitae	Squash mosaic virus
Hyperparasitism	Many	various fungi	Trichoderma spp.
	Sunflower, Beans	Sclerotinia spp.	Coniothyrium minitans
	Lettuce	Sclerotinia spp.	Sporodesmium sclerotivorum
	Sugar beet	Pythium spp.	Pythium oligandrum
	Cucumber, Beans	Rhizoctonia solani	Laetisaria arvalis
	Cucumber	mildews	Ampelomyces grisqualis
	Rye, other cereals	Ergot	Fusarium roseum 'heterosporum'
Hypovirulence	Chestnut	Endothia parasitica	Mycovirus
Parasitism	Soybean	Pseudomonas syringae pv. glycinea	Bdellovibrio bacteriovorus
Predation		Various fungi	Arachnula impatiens

(From Schroth & Hancock, 1985.)

be patented although their applications can. A problem for the
future concerns the use of genetically engineered strains of
antagonists.

Should the emphasis in agriculture shift from intensive to lower
input, less intensive or organic farming, and the current trend for
increasing resistance to fungicides continue, then perhaps the era of
biological control will dawn.

Forecasting disease

According to Tait (1987), forecasting plant disease is an attempt to
estimate the future state of disease in a crop from observations of
the current or recent state of disease or from measurements of
related factors. A thorough understanding of the life-cycle and
epidemiology of a plant pathogen and its relationship to other
diseases and pests in a given crop system is essential for a reliable
forecast. Reliable disease forecasting should provide the farmer
with a more efficient, rational basis for his decisions about control.

Observations of the state of disease include monitoring levels of
disease in a crop as outlined in 'disease risk assessment' (chemical
control). The principle of disease risk assessment is that fungicides
are applied only when there is a risk of disease becoming severe
enough to cause yield loss. For example, estimates of stem base
browning in cereals caused by the eyespot fungus *Pseudocercosporella herpotrichoides* during early spring may form the basis for a
decision to apply a fungicide. This procedure should be reliable
because most infections by eyespot occur early in the season and the
pathogen is largely monocyclic. In addition, spores only spread over
limited distances so there is unlikely to be any extraneous inoculum.
In many diseases, for example powdery mildew, extraneous
inoculum may enter the crop resulting in higher levels of disease
than expected. The main limitation of this procedure with eyespot is
that the severity of eyespot lesions cannot be predicted. Factors
leading to the development of severe eyespot lesions late on in the
season, which cause yield loss, are not well understood. In some
diseases, particularly those caused by soil-borne pathogens, the
amount of initial inoculum plays a major role in the development of
disease. Forecasts can therefore be made based on direct or indirect
assessments of initial inoculum. For example it is possible to count
directly certain fungal resting bodies in soil, e.g. sclerotia. Soils
containing large numbers of resting bodies should be avoided for
susceptible crops. Alternatively, levels of soil-borne pathogens may
be estimated indirectly by planting susceptible hosts in soil samples

and growing plants in a glasshouse. Soils that produce most severe disease should be avoided. Such tests are used to predict levels of clubroot using oilseed rape as the test host.

It is evident that disease forecasting is no easy task and many factors relating to the host, pathogen and environment may have to be involved in the production of a forecasting 'model'. Perhaps the most common 'related factors' concern the environment. Weather conditions during the inter-crop period as well as during the crop season may influence opportunities to forecast disease. Extreme weather conditions between crops clearly affect the survival of a pathogen or its vectors. According to Tarr (1972), the severity of blue mould of tobacco (caused by *Peronospora tabacina*) in the southern USA appeared to be closely related to winter temperatures. Mild January temperatures led to more severe attacks, and low temperatures in January resulted in late appearance of blue mould and relatively little damage even when conditions during the growing season were favourable for its spread. A further example concerns bacterial wilt of maize caused by *Erwinia stewartii* in the USA. The primary vector of this bacterium is a flea beetle which dies during severe winters. Hence, if the sum of the mean monthly temperatures for December to February exceeds 32 °C, wilt occurs; if the temperature exceeds 37.8 °C, wilt will probably be severe. Generally farmers assume that many diseases are most severe after a mild humid winter.

The commonest attempts at forecasting disease outbreaks are based on weather conditions during the crop season. To reconsider briefly the effect of the environment on disease development (Chapter 1), the major influences are often temperature and moisture (rain or humidity). These form the basis for a number of forecasting rules including those for late blight of potatoes, downy mildew of hops, scab of apples, and *Septoria* infection of cereals. However, such systems do not as yet incorporate weather forecasts but interpret past weather data and the time interval given by the latent period of the pathogen to predict when symptoms will appear or sporulation occur. Water as daily rainfall, rainfall intensity and duration, humidity and leaf surface wetness together with temperature have been incorporated into a considerable range of disease forecast models for a variety of crops (Royle & Butler, 1986). Individual models will be discussed under specific crop diseases.

At first sight, it appears that disease forecasting should indeed be one of the keys to rational decision-making in crop protection programmes. Some of the models have been thoroughly researched and others have been developed over many years. In the UK, for example, the Smith Period for forecasting potato blight is the latest refinement in a succession of forecasting models which date back over 55 years to the four 'Dutch rules', the first successful warning system for potato blight, developed in Holland. A Smith Period is defined as two consecutive days ending at 0900 hours when the temperature was not less than 10 °C and the relative humidity above 90% for at least 11 hours of each day. This would not be beyond the farmer to monitor himself, but the ADAS in the UK routinely warns farmers of the occurrence of Smith Periods in their locality through farming programmes on radio and TV, on 'Prestel', in the farming press, on ADAS bulletins and on telephone information services. In the USA a computerised forecast for potato blight called BLITECAST has been developed (Fry, 1987). Blight forecast information in the UK is readily available, and cheap (or free). However, Royle & Shaw (1988) state that 'fully operational forecasting systems that guide tactical decisions about control measures on economic or other grounds are very scarce'. Generally, farmers do not use disease forecasting methods. Royle and Shaw suggest two major reasons why.

The first concerns the reliability of forecasting methods. Control decisions based on forecasting have to be financially safe; the cost of failing to control disease can be very high. In the case of potato blight, a disease which, if unchecked, can devastate a highly valuable crop in as little as a week, most farmers prefer to rely on a routine prophylactic spray programme. In addition, in order to improve the accuracy of disease forecasting, weather conditions should ideally be monitored by the farmer on his farm, possibly even in each field. The time and effort involved would probably be prohibitive. There are now available automatic electronic weather stations which, when linked to a suitably programmed microcomputer, can advise the farmer on conditions likely to favour outbreaks of disease on the farm, although these would cost several thousand pounds to set up. Hence the second main reason for the unpopularity of disease forecasting is that the procedures have been judged by farmers to be of greater average cost than average benefit. There are few examples of cost–benefit analysis of disease forecasting versus routine prophylactic spraying, and until the farmer is convinced otherwise, he is likely to continue with his own tried and tested disease control strategies.

There may, however, be positive benefits of disease forecasting. If an awareness of forecasting, initially at least, motivates the farmer to identify and monitor levels of disease correctly, the emphasis on fungicide control may shift from a routine prophylactic programme to one where disease thresholds are involved and fewer fungicide sprays used. According to Wilson (1987), farmers using EPIPRE, a computerised system of supervised control of diseases and pests in winter wheat have been 'keen to learn and prepared to seek a greater understanding of their crops'.

Integrated control

Integrated control of diseases or integrated pest/pathogen management (IPM) is a most topical subject. For many diseases no single method of control is totally satisfactory. A combination of all the available methods of control may be required in order to reduce the incidence of disease to acceptable levels. Even if one method of control, e.g. fungicides, is effective, farmers may be advised to adopt a strategy of integrated control in order to reduce the risk of losing the efficacy of that single method. Integrated control is simply the continuous application of a balanced range of disease control measures: a commonsense and economical approach to disease control. The main components of an integrated control programme are cultural control, resistant varieties and fungicides; legislative control, disease forecasting and biological control should be used where available. The challenge of an integrated control strategy is to balance all the components correctly in a programme which is flexible enough to allow for unexpected disease epidemics. This balance will probably vary from field to field and year to year. Hence, the best integrated control programmes are probably devised after many years' experience and many be specific to individual fields. In addition, a farmer will also have to consider problems of pests and weeds in his integrated control strategy. Much time, effort and expense can be saved if, for example, suitable herbicides, pesticides and fungicides can all be applied at the same time as a tank mixture.

In order to illustrate some of the decisions a farmer must make in an integrated control strategy, an example of an integrated control programme is given below for winter wheat.

Pre-sowing

Select varieties. Yield and quality are probably the prime consideration, but consider also resistance to diseases which are a problem locally. If growing more than one wheat variety, select a range of varieties, as far as possible according to the winter wheat diversification scheme. In low-input or organic farming consider variety mixtures or blends. Use 'certified seed' to reduce seed-borne disease problems. Cereal seed will most probably already be treated with a fungicide. In intensive cereal situations, where crops are early-drilled and may suffer from take-all and other early-season disease problems, it is probably wise to use one of the 'systemic' seed treatments. Otherwise, a cheap 'non-systemic' seed treatment should suffice. Rotations and crop husbandry are important in disease control: where possible, no more than two successive cereal crops should be grown. With regard to crop husbandry, any debris from previous crops should be disposed of properly by burning or ploughing. The field should also be free of weeds and volunteers. Cultivations should produce a firm seedbed and good tilth.

At sowing
Sow at optimum time (around beginning of October in the Midlands, UK). Early-drilled crops suffer from higher levels of disease. Late-drilled crops may have problems if the seedlings emerge during freezing winter weather. Use optimum seed rates. Densely packed plants increase humidity under crop canopy and encourage diseases such as mildew.

Autumn and winter
Generally no fungicide treatment should be required, particularly if a systemic seed treatment has been used. It is advisable, however, to examine the crop periodically and assess levels of mildew, *Septoria tritici* and stem base browning (eyespot, sharp eyespot, *Fusarium*), which may require treatment in early spring. In the autumn (October/November), it may be necessary to apply an aphicide to protect the crop against barley yellow dwarf virus. Factors which increase the likelihood of BYDV infection include sowing after grass, early sowing and sowing in a high risk area with a history of BYDV infection.

March/April (growth stage (GS) 30–33)
The most important disease to control at this stage is eyespot. Assess the crop for extent of stem base browning (assumed to be eyespot). If more than 20% tillers show stem base browning, apply a suitable fungicide. Some farmers, particularly those growing continuous cereals, will apply a spray to control eyespot at this stage irrespective of disease levels. Other diseases which may be troublesome are mildew and *Septoria tritici*. If there are signs of these diseases building up in the crop, apply a fungicide or mixture of fungicides which are appropriate for all diseases (including eyespot). Mid to late April is also the best time for the main

application of nitrogen fertiliser. Applications prior to this (mid-March) and subsequently (early June) may also be required. Ensure that optimum amounts of fertiliser are used. Excess nitrogen encourages diseases such as mildew.

May (GS 33–37)

If a broad-spectrum fungicide has been applied at GS 30–33, it is unlikely that a further application will be required during May. However, if no fungicide was applied, the crop should be carefully monitored for disease. Mildew and *Septoria* spp. are likely to be the main problem, although some yellow rust may be just starting to develop. Do not allow disease to exceed threshold levels (see cereal diseases). Apply fungicide appropriate to diseases present. Also check again for stem base browning. It is still possible to reduce the effects of eyespot by applying a fungicide up to GS 37.

Late May–mid June (GS 39–55)

Fungicide treatment now is likely to be necessary if the first spray was at GS 30–33, but not if it was at GS 33–37. This period around flag leaf emergence is very important as far as foliar diseases are concerned. Perhaps the most important diseases are the *Septoria* spp. (*S. nodorum* and *S. tritici*) which can kill large areas of flag leaf. *Septoria* is a serious problem particularly after periods of wet weather. If in the 2 weeks previous to flag leaf emergence there have been 3 consecutive days with at least 10 mm of rain, it is wise to apply a fungicide to protect the flag leaf. Other diseases such as yellow and brown rust and mildew may also be a problem, especially if the variety is susceptible to these diseases. It may be necessary, therefore, to apply a broad-spectrum fungicide with activity against all foliar diseases.

End of June, beginning of July (GS 59)

A fungicide applied at GS 39–55 reduces the need for an application at this stage. However, if the last time a fungicide was applied was around GS 33, then consider a further application, based on the same criteria as for the GS 39–55 application.

July (GS 60–71)

This final spray, aimed at protecting the ear and reducing any further late-season diseases on the flag leaf, may not be necessary if a fungicide has been applied at GS 59. However, diseases such as glume blotch (*S. nodorum*) and *Fusarium* ear blight can be troublesome on the wheat ear, particularly during wet seasons. Ears may also become infected by mildew and rust during severe disease epidemics. Quality and yield of wheat can be dramatically reduced by severe ear infections and therefore a late-season fungicide spray may be necessary, particularly if disease is developing on the flag leaf and there has recently been a period of wet weather or rain is imminent. Generally, sprays should not be applied after the grain has reached the milky ripe stage.

It is important to note that in a programme where a number of fungicides may be applied the choice of fungicides is very important. Not only should fungicides be approved for the use that they are intended, but when more than one spray is used, a diversity of fungicides with different modes of action should be selected. This will avoid fungicide-resistant stains of pathogens building up and increase the spectrum of diseases controlled. Crop monitoring and non-chemical control should not be neglected. It would not be impossible (although unfortunately unlikely) for a wheat crop to go through a season without disease thresholds being reached and fungicide application being deemed unnecessary.

Further reading

Legislative control; cultural control

Yarham, D. J. (1988). The contribution and value of cultural practices to control arable crop diseases. In *Control of Plant Diseases: Costs and Benefits*, ed. B. C. Clifford & E. Lester, pp. 135–54. Oxford: Blackwell.

Yarham, D. J. & Norton, J. (1981). Effects of cultivation methods on disease. In *Strategies for the Control of Cereal Disease*, ed. J. F. Jenkyn & R. T. Plumb, pp. 157–66. Oxford: Blackwell.

Disease resistance

Bailey, J. A. & Mansfield, J. W. (ed.) (1982). *Phytoalexins*. Glasgow and London: Blackie.

Clifford, B. C. & Lester, E. (ed.) (1988). *Control of Plant Diseases: Costs and Benefits*. Oxford: Blackwell.

Vanderplank, J. E. (1963). *Plant Diseases: Epidemics and Control*. New York: Academic Press.

Vanderplank, J. E. (1984). *Disease Resistance in Plants*. New York: Academic Press.

Wood, R. K. S. (1967). *Physiological Plant Pathology*. Oxford: Blackwell.

Chemical control

Ainsworth, G. C. (1981). *Introduction to the History of Plant Pathology*. Cambridge: Cambridge University Press.

Brent, K. (1985). One hundred years of fungicide use. In *Fungicides for Crop Protection, BCPC Monograph 31*, ed. I. M. Smith, pp. 11–22. Thornton Heath, Surrey: BCPC publication.

Brent, K. J. & Atkin, R. K. (ed.) (1987). *Rational Pesticide Use*. Cambridge: Cambridge University Press.

Buchenauer, H. (1985). General review of the mode of action of fungicides. In *Fungicides for Crop Protection, BCPC Monograph 31*, ed. I. M. Smith, pp. 55–65. Thornton Heath, Surrey: BCPC publication.

Clifford, B. C. & Lester, E. (ed.) (1988). *Control of Plant Diseases: Costs and Benefits*. Oxford: Blackwell.

Ford, M. G., Hollomon, D. W., Khambay, B. P. S. & Sawicki, R. M. (ed.) (1987). *Combating Resistance to Xenobiotics: Biological and Chemical Approaches*. Chichester: Ellis Horwood.

Hassall, K. A. (1982). *The Chemistry of Pesticides*. London: Macmillan.

Holloman, D. W. (1986). Contribution of fundamental research to combating resistance. In *Proceedings of the 1986 British Crop Protection Conference – Pests and Diseases*, pp. 801–10. Thornton Heath, Surrey: BCPC publication.

Ivens, G. W. (ed.) (1989). *The UK Pesticide Guide*. CAB and BCPC publication.

Locke, T. (1986). Current incidence of fungicide resistance in pathogens of cereals in the UK. In *Proceedings of the 1986 British Crop Protection Conference – Pests and Diseases*, pp. 781–86. Thornton Heath, Surrey: BCPC publication.

Lyr, H. (ed.) (1987). *Modern Selective Fungicides – Properties, Applications, Mechanisms of Action*. Harlow: Longman.

Matthews, G. A. (1979). *Pesticide Application Methods*. London: Longman.

Ministry of Agriculture, Fisheries and Food (1986). *Use of Fungicides and Insecticides on Cereals*. Booklet 2257 (86). Alnwick, Northumberland: MAFF Publications.

Scopes, N. & Ledieu, M. (ed.) (1983). *Pest and Disease Control Handbook*. Thornton Heath, Surrey: BCPC publication.

Shephard, M. C. (1981). Factors which influence the biological performance of pesticides. In *Proceedings of the 1981 British Crop Protection Conference – Pests and Diseases*, pp. 711–21. Thornton Heath, Surrey: BCPC publication.

Shephard, M. C. (1985). Fungicide behaviour in the plant-systemicity. In *Fungicides for Crop Protection, BCPC Monograph 31*, ed. I. M. Smith, pp. 99–106. Thornton Heath, Surrey: BCPC publication.

Biological control

Baker, K. F. (1987). Evolving concepts of biological control of plant pathogens. *Annual Review of Phytopathology*, **25**, 67–85.

Baker, K. F. & Cook, R. J. (1982). *Biological Control of Plant Pathogens*. St. Paul: The American Phytopathological Society.

Cook, R. J. & Baker, K. F. (1983). *The Nature and Practice of Biological Control of Plant Pathogens*. St Paul: The American Phytopathological Society.

Deacon, J. W. (1983). *Microbial Control of Plant Pests and Diseases. Aspects of Microbiology 7*. Wokingham: Van Nostrand Reinhold (UK).

Hoy, M. A. & Herzog, D. C. (ed.) (1985). *Biological Control in Agricultural IPM Systems*. Florida: Academic Press.

Forecasting disease

Austin, R. B. (1982). *Decision Making in the Practice of Crop Protection, BCPC Monograph 25*. Thornton Heath, Surrey: BCPC publication.

Butt, D. J. & Jeger, M. J. (1985). The practical implementation of models in crop disease management. In *Advances in Plant Pathology, Vol. 3, Mathematical Modelling of Crop Disease*, ed. C. A. Gilligan, pp. 207–30. London: Academic Press.

Fry, W. E. (1987). Advances in disease forecasting. In *Rational Pesticide Use*, ed. K. J. Brent & R. K. Atkin, pp. 239–52. Cambridge: Cambridge University Press.

Polley, R. W. & Clarkson, J. D. S. (1978). Forecasting cereal disease epidemics. In *Plant Disease Epidemiology*, ed. P. R. Scott & A. Bainbridge, pp. 141–50. Oxford: Blackwell.

Royle, D. J. & Butler, D. R. (1986). Epidemiological significance of liquid water in crop canopies and its role in disease forecasting. In *Water, Fungi and Plants*, ed. P. G. Ayres & L. Boddy, pp. 139–56. Cambridge: Cambridge University Press.

Royle, D. J. & Shaw, M. W. (1988). The costs and benefits of disease forecasting in farming practice. In *Control of Plant Diseases: Costs and Benefits*, ed. B. C. Clifford & E. Lester, pp. 231–46. Oxford: Blackwell.

Tait, E. J. (1987). Rationality in pesticide use and the role of forecasting. In *Rational Pesticide Use*, ed. K. J. Brent & R. K. Atkin, pp. 225–38. Cambridge: Cambridge University Press.

Wilson, J. (1987). Commercial implementation of forecast methods. In *Rational Pesticide Use*, ed. K. J. Brent & R. K. Atkin, pp. 333–41. Cambridge: Cambridge University Press.

Integrated control

Ministry of Agriculture, Fisheries and Food (1986). *Use of Fungicides and Insecticides on Cereals*. ADAS Booklet 2257 (86). Alnwick, Northumberland: MAFF Publications.

II PRACTICE

Introduction

The second section of the book describes the application of plant pathology to individual crops. A catalogue of diseases is presented which includes all those which are widespread and/or of economic importance on all the major temperate crops.

For each disease–host combination, the causal organism, host range, symptoms (and, where possible, assessment), disease cycle, significance and control will be described. Aspects such as recognition, significance and control of plant diseases have obvious direct practical applications to agriculture. In addition, an understanding of the host range is important in devising crop rotations. Finally, life cycles of pathogens, including information on reproduction, dispersal, epidemiology and survival, have obvious practical implications with regard to control of plant diseases.

All chemicals recommended for control of disease have been taken from *The UK Pesticide Guide* (Ivens, 1989), except for soybean diseases where the *Crop Protection Chemicals Reference, 4th Edition, 1988* has been used. Product manuals should be consulted for further information on approved uses and methods of application.

6 Diseases of small-grain cereals

Despite fluctuations in world market prices and European Community (EC) subsidies, wheat, barley and oats are still the most important crops in many temperate areas of the world, including the UK. It is well understood that the yield of cereals can be adversely affected by disease and as a result of much effort by plant breeders, agrochemical companies, agronomists and plant pathologists, some highly effective control measures have been developed. The effect of diseases on the quality of cereals is less well understood but of increasing importance in markets where premium prices are paid for breadmaking wheat, and barley which can be used in the brewing industry (malting barley).

Disease may affect all parts of a cereal plant, i.e. roots, stem base, stem, leaves and ears, and the major diseases affecting each part will now be discussed.

Take-all

(i) *Causal organism*
Fungus, *Gaeumannomyces graminis*.

(ii) *Host range*
Gaeumannomyces graminis var. *tritici* attacks wheat, barley, rye and many grass species. Oats are immune. *Gaeumannomyces graminis* var. *avenae* attacks oats, wheat, barley, rye and many grass species.

(iii) *Symptoms (Figure 6.1)*
The first symptoms of take-all in autumn-sown crops may be seen in the late autumn or early spring. Patches of yellow or stunted

seedlings become apparent and the reduction in competitiveness of the crop may lead to weed infestation. More frequently, symptoms are seen near to maturity, when patches of stunted plants and bleached heads (whiteheads) appear, particularly after a dry period. Whiteheads generally contain small grains or, occasionally, no grain. If diseased plants are pulled up, there is often little resistance from the roots which have become blackened and stunted. Some plants may have developed 'buttress roots' from their first node in order to compensate for the destruction of the

Figure 6.1. Take-all of cereals caused by *Gaeumannomyces graminis*. Healthy plants on left. Note stunted diseased plants with blackened stunted roots and stem bases. (J. D. S. Clarkson, ADAS.)

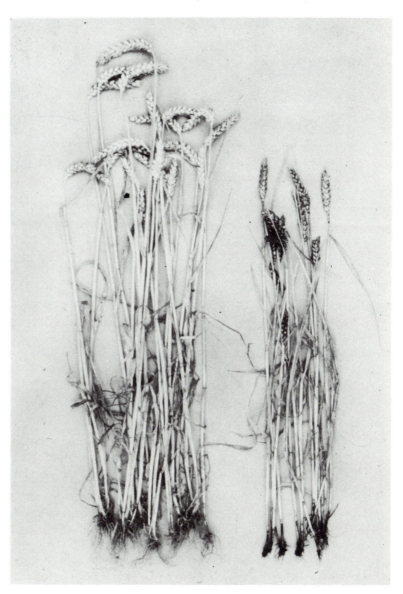

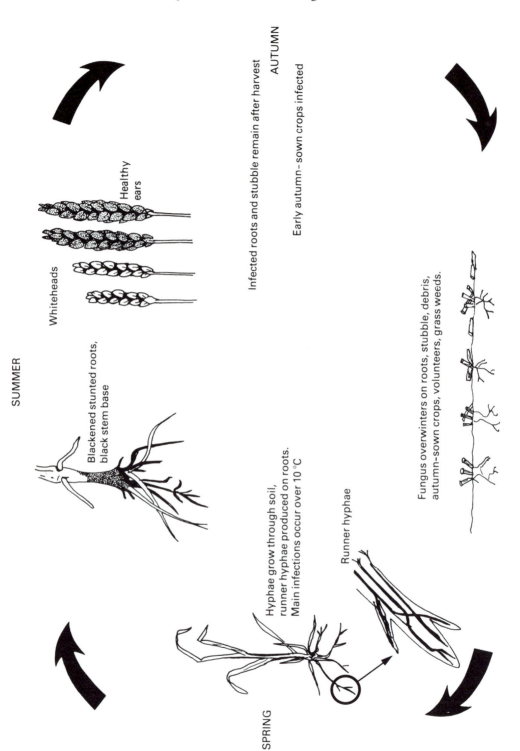

AUTUMN

Infected roots and stubble remain after harvest

Early autumn-sown crops infected

Healthy ears

Whiteheads

SUMMER

Blackened stunted roots, black stem base

WINTER

Fungus overwinters on roots, stubble, debris, autumn-sown crops, volunteers, grass weeds.

Hyphae grow through soil, runner hyphae produced on roots. Main infections occur over 10°C

Runner hyphae

SPRING

main roots. The base of infected plants may also show blackening with dark mycelium and, particularly during wet weather, perithecia may be seen as small dark raised spots.

(iv) *Disease cycle (Figure 6.2)*

Take-all inoculum survives the winter primarily as mycelium on roots or stubble debris. Volunteer cereals, early autumn-sown crops and some rhizomatous grass weeds may also act as overwintering sources of inoculum. The main infections of cereal roots occur from late May onwards in the UK when temperatures exceed 10 °C. However, autumn infection can occur in early-sown crops during mild weather. Mycelium grows from its food source onto cereal roots and the fungus then spreads along the roots as long dark runner hyphae. Periodically along the runner hyphae, loose gatherings of hyphae called hyphopodia occur. Infection pegs which penetrate the root develop underneath the hyphopodia. Although both the root endodermis and vascular tissue may be colonised by *G. graminis*, the fungus does not grow systemically in the plant. As the root rot progresses during the growing season, the ability of the plant to absorb water and nutrients from the soil declines. If there is sufficient reduction in root absorbing capacity, the cereal ears are deprived of water and nutrients and hence whiteheads develop. Ascospores produced from perithecia of *G. graminis* have also been shown to be capable of infecting roots exposed above the soil surface, but they are considered unimportant under normal field conditions. Biological control by soil micro-organisms may be the main reason for the poor germination and growth of ascospores in soil. Ascospores can be important in introducing the fungus into reclaimed land (e.g. the Dutch polders) where natural antagonists are absent. After harvest, the fungus remains in the roots and stubble, and will be viable until the residues rot away. Take-all is most serious in light loose alkaline soils but severe attacks can also occur in acid patches. Badly drained fields and poor nutrient status also encourage the disease. These latter factors are probably related to poor root growth in the host rather than direct encouragement of the pathogen. Take-all is also encouraged by early sowing, puffy seedbeds and continuous cereal cultivation, until take-all decline occurs (see Control).

(v) *Significance*

Figure 6.2. Disease cycle of take-all (*Gaeumannomyces graminis*).

Take-all is probably the most significant disease of cereals in temperate areas; it occurs almost everywhere that cereals are grown and can dramatically reduce yields. It is very difficult to estimate

yield losses as they vary considerably from year to year. Polley &
Clarkson (1980) calculated that moderate and severe levels of
take-all reduced yields of individual tillers by 16 and 62%,
respectively. In a farm crop, during second and third successive
wheat crops, yield losses are likely to be in the order of 10–20%.
Yield losses may go unnoticed if symptoms are not obvious. Clearly
take-all can limit cereal production. Severe take-all can also
significantly reduce 1000-grain weight and specific weight of grain.

(vi) *Control*

Cultural control offers the best method of reducing take-all. A
one-year break from susceptible crops will almost eliminate the
problem for the subsequent cereal crop unless take-all decline has
been reached (see below). Other cultural practices can also
influence the disease. Seed should be drilled at an optimum time;
early drilled crops can suffer severe disease attacks. It is advisable to
sow first wheat crops first, long-term wheats next and second and
third wheats last. Seedbeds should be firm and the soil well drained
with any structural faults rectified, particularly in light soils. A
balanced fertiliser programme especially with regard to nitrogen,
phosphate and manganese should be adopted to ensure good root
growth, and volunteers together with grass weeds should be
eliminated. There is also some evidence to suggest that ploughing in
straw rather than incorporating it by minimal tillage techniques may
also reduce the disease.

Both cultural control and biological control are probably
involved in the phenomenon of take-all decline. Should a farmer
choose to grow successive cereals, take-all will generally build up
during the second to fourth years. However, take-all will almost
invariably decline during subsequent years until it reaches a
base-line which may allow the farmer to continue growing the crop
economically (Figure 6.3). The reasons for this rather curious
phenomenon are a matter for much debate. It is possible that the
suppression in the soil is the result of a build-up of micro-organisms
antagonistic towards *G. graminis*. Another theory suggests that *G.
graminis* becomes avirulent possibly as a result of virus infection.
Take-all decline may be used as a means of control especially on
those soils where yield losses are not particularly severe and the
profitability of continuous cereal exceeds that of a cereal/break
system. If the run of continuous cereals is broken by a non-
susceptible crop, the whole process will be repeated with subsequent cereal crops.

Disease resistance in varieties of wheat, barley and oats to the respective subspecies of *G. graminis* is disappointingly low. However, *G. graminis* var. *avenae* is not widespread in the UK and many farmers exploit the resistance of oats to the more common *G. graminis* var. *tritici* by using oats as a break crop. There is also evidence of quite good levels of disease resistance in some triticale varieties. Triticale (a cross between wheat and rye) showed intermediate levels of resistance between wheat (susceptible) and rye (resistant) (Hollins, Scott & Gregory, 1986). Growing the more resistant varieties of triticale may be a further means of introducing take-all resistance into cereal cultivations.

Chemical control of take-all in the field is still not feasible on a commercial scale. There have been many attempts to develop chemical control methods but most have met with little success as far as the farmer is concerned. Treating seed with triadimenol + fuberidazole (Baytan) may suppress the disease, particularly if cereals are early-sown in a high risk situation. Other possible methods of chemical control under investigation include the application of fungicides, such as the benzimidazoles and triazoles, as granules or pellets with the seed at planting time (Ballinger & Kollmorgen, 1986). Soil drenches with fungicides such as nuarimol have also shown promise in experimental work (Bateman, 1984). However, at the rates required for effective control, such treatments are prohibitively expensive.

Biological control of take-all is an area of considerable interest, mainly because of the lack of any other method of control in continuous cereals. In addition, the phenomenon of take-all decline

Figure 6.3. Diagrammatic representation of take-all decline. ———, take-all infection; – – –, yield.

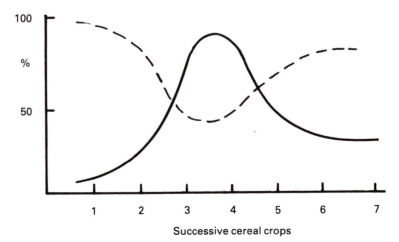

Successive cereal crops

is widely understood to occur as the result of some type of biological control, and numerous attempts have been made to isolate the organism(s) [antagonist(s)] responsible for the eventual demise of take-all in the field. Fungal species of the genus *Phialophora* have shown some capacity to reduce the effects of take-all when inoculum is introduced into take-all infested soil. The development of this has yet to be exploited on a commercial scale. However, there are moves in the USA to introduce another antagonist to take-all, *Pseudomonas fluorescens*, into commercial agriculture. When used as a seed treatment the bacterium appears to offer some protection against take-all and yield increases of up to 15% have been observed in the USA. Unfortunately the results in the UK have been less satisfactory but work on this and other antagonists continues.

Integrated control of take-all using a combination of all the cultural control methods available, together with a systemic seed treatment in certain circumstances, is the most effective way of reducing this potentially devastating disease.

Eyespot

(i) *Causal organism*
Fungus, *Pseudocercosporella herpotrichoides*.

(ii) *Host range*
Wheat, barley, oats, rye and many grass species, primarily winter wheat and winter barley.

(iii) *Symptoms (Figure 6.4)*
In late autumn and early spring, eyespot symptoms may be difficult to distinguish from those of the other stem base pathogens, sharp eyespot and *Fusarium*. All that is visible is a brown smudge on the leaf sheath at the stem base. Sometimes, particularly if the crops were very early-sown, an eyespot lesion may penetrate one or two leaf sheaths and small, almost black pinpricks may be seen on the white background of the leaf sheath. Very occasionally the 'eyespot' lesion, characteristic of symptoms in mature plants, can just be seen starting to form. Eyespot symptoms in mature plants consist of a dark brown, eye-shaped lesion which is generally below the first node of the plant. The lesions have a diffuse margin (cf. sharp eyespot) and a central black 'pupil' consisting of a mass of compacted hyphae. If stems of colonised plants are broken open, a grey mycelium may be found in the cavity. Severe eyespot lesions

which penetrate the stem can result in breakage and 'straggling' (a few tillers falling down) or 'lodging' (many tillers falling down). Whiteheads may also be produced, giving similar symptoms to take-all. An assessment key for eyespot based on three categories of disease (slight, moderate and severe) is given in Chapter 3.

(iv) *Disease cycle (Figure 6.5)*

The eyespot pathogen overwinters primarily on infected stubble. Early-sown autumn crops, volunteers and some grass weeds may also act as sources of inoculum. Although *P. herpotrichoides* is poorly competitive in soil, it can survive on stubble for as long as 3 years. Conidia of the fungus may be produced at any time during the late autumn and winter, but infections occur at temperatures above 5 °C and during wet periods. Conidia are splashed short distances, from infected stubble to healthy stem bases, by rain droplets. Early infections generally allow time for the fungus to penetrate

Figure 6.4. Eyespot of cereals caused by *Pseudocercosporella herpotrichoides*. (R. Gutteridge, Rothamsted.)

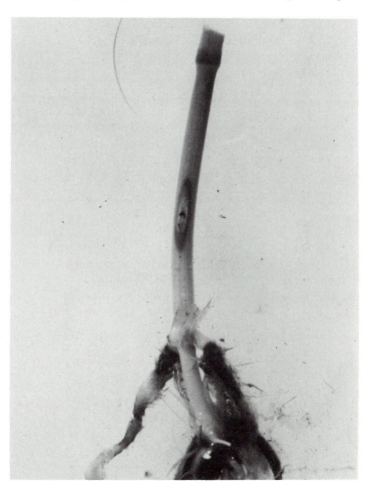

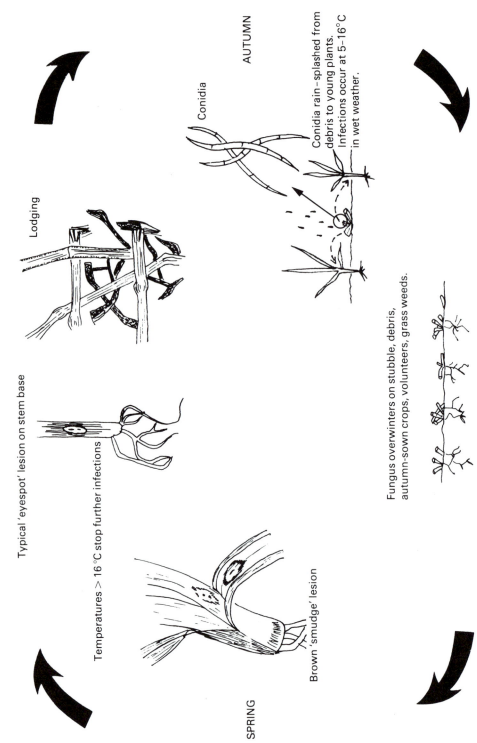

AUTUMN

Conidia

Conidia rain-splashed from debris to young plants. Infections occur at 5–16°C in wet weather.

Lodging

SUMMER

Typical 'eyespot' lesion on stem base

Temperatures > 16°C stop further infections

Fungus overwinters on stubble, debris, autumn-sown crops, volunteers, grass weeds.

WINTER

Brown 'smudge' lesion

SPRING

successively all the leaf sheaths and cause severe lesions on the stem. The infection process may take 8 weeks, depending upon environmental conditions. The disease is generally understood to be monocyclic, although during mild damp winters or early springs it is possible for conidia to be produced on an infected plant and dispersed in the crop causing secondary infections. Fresh infections are unlikely to occur at temperatures above 16°C. After harvest cereal stubble remains infected and could provide additional inoculum for successive crops. Hence eyespot tends to be more of a problem in continuous cereals, where inoculum may build up from year to year. Heavy wet clay soils are also conducive to eyespot as is excessive nitrogen fertiliser usage. The latter encourages tillering, which results in a dense and humid crop canopy and also promotes stem elongation, which makes plants additionally prone to lodging. As previously mentioned, early-sown crops are at risk from severe eyespot attacks.

(v) *Significance*

It is generally accepted that slight eyespot does not reduce yield significantly. However, moderate and severe eyespot has been seen to reduce yield on individual tillers by 10 and 36%, respectively (Clarkson, 1981). On a field scale, losses from even the most severe attacks are unlikely to exceed 50% and in winter cereals losses probably average 5–10%. Lodging caused by eyespot slows down the harvesting process and can result in poorer quality grain of high moisture content. The disease has recenty taken on a new significance because of the widespread distribution of MBC-resistant strains.

(vi) *Control*

Cultural control methods include optimum sowing time, avoiding excess nitrogen fertiliser, disposal of infected stubble, and a two-year break from cereals. Lodging in mature plants may be reduced by the application of a growth regulator such as cycocel.

Disease resistance in winter wheat for many years in the UK has been derived mainly from the variety Cappelle Desprez which confers on varieties a degree of tolerance rather than total resistance to disease. Indeed, it is unlikely that plant breeders will be able to produce a commercial variety which is totally resistant to disease, but neither is this essential. Slight attacks of eyespot, as previously discussed, do not affect yield significantly. Recently, however, there has been a breakthrough in resistance to eyespot with the introduction of the winter wheat variety Rendezvous onto

Figure 6.5. Disease cycle of eyespot (*Pseudocercosporella herpotrichoides*).

the National Institute of Agricultural Botany's (NIAB) Farmers' Leaflet No. 8, *Recommended Varieties of Cereals*. This variety shows a much higher degree of resistance to eyespot than any other popular variety (Hollins *et al.*, 1988), although under maximum eyespot pressure, severe lesions may still occur. The resistance of Rendezvous has been derived from a grass species, *Aegilops ventricosa*. It is likely that other varieties will also eventually exploit the same source of eyespot resistance.

Chemical control. Until the early 1980s, application of MBC (methyl benzimidazole-2-yl carbamate) fungicides such as carbendazim or benomyl gave good control. Total eradication of eyespot is difficult and fungicides are aimed at suppressing the development of severe eyespot lesions. From 1974 to 1982 there was an overall increase in the use of fungicides on cereals in the UK and over 50% of winter wheat crops were treated with at least one application of an MBC compound in 1982. However, during 1981 the first reports of failure to control eyespot using MBC fungicides occurred. Eyespot had started to develop resistance to MBC fungicides. The problem rapidly became widespread in the UK and it is now estimated on the basis of national surveys that MBC-resistant eyespot isolates exist in most winter cereal crops. The efficacy and comparative cheapness of MBC fungicides may well have hastened their downfall. Farmers used MBCs extensively, even when the threat of eyespot was not great. Thus the selection pressure on the pathogen to produce resistant isolates was increased.

There are two main types of isolates of *P. herpotrichoides*, the wheat (W) and rye (R) types. W-types are more pathogenic on wheat and barley than rye and *R*-types are pathogenic on wheat, barley and rye. In culture, W-types are faster growing and have an even margin; R-types are slower growing and have a feathery edge to the colony. A comprehensive survey of morphology, MBC resistance and pathogenicity of eyespot isolates by Hollins, Scott & Paine (1985) showed that there had been a significant increase in the number of R-type isolates found since 1981 in areas with no history of rye growing. This population shift coincided with an increase in the proportion of MBC-resistant isolates in the population. Hence it was suggested that MBCs select simultaneously for resistance and the R-type isolates by preferentially controlling the W-type isolates in the eyespot population. The selection for R-type isolates may also be significant as far as winter barley is concerned: winter barley seems to be more susceptible to R-type isolates than winter wheat.

Chemicals approved for use against eyespot are given below, together with some examples of proprietary products.

Active ingredient	Product example
benomyl	Benlate
carbendazim	Bavistin, Derosal Liquid, Focal Flowable, Stempor DG
carbendazim + chlorothalonil	Bravocarb
carbendazim + flutriafol	Early Impact
carbendazim + mancozeb	Kombat
carbendazim + maneb	Multi-W FL
carbendazim + maneb + sulphur	Bolda FL
carbendazim + maneb + tridemorph	Cosmic FL
carbendazim + prochloraz	Sportak Alpha
carbendazim + propiconazole	Hispor 45 WP
maneb	Manzate
prochloraz	Sportak
propiconazole	Radar, Tilt
thiophanate-methyl	Cercobin

Although the list is quite extensive, the majority of approved chemicals contain MBC fungicides and cannot be relied upon to give adequate control. The ADAS recommended treatment for control of eyespot is Sportak (prochloraz) on winter wheat and Sportak Alpha (carbendazim + prochloraz) on winter barley. As yet, there is no evidence of prochloraz resistance in the eyespot population. The majority of the products listed should be applied quite early on in the growing season, at GS 31, in order to stop penetration of leaf sheaths by the fungus. Some farmers, particularly in intensive or continuous cereal situations, consider that a fungicide application at this growth stage is essential for eyespot control and spray prophylactically regardless of disease levels. Such prophylactic treatment is undoubtedly the main reason for development of MBC resistance. In situations where the risk of eyespot is not so great, it may be more cost-effective to apply a fungicide only when a disease threshold is reached. A disease threshold of 20% tillers affected is recommended by ADAS.

Integrated control using cultural methods, resistant varieties and, if necessary, fungicides is the best way of reducing the threat of eyespot in winter cereals.

Sharp eyespot

(i) *Causal organism*
Fungus, *Rhizoctonia cerealis* (*Ceratobasidium cereale*).

(ii) *Host range*
Wheat, barley, oats, rye and some grass species.

(iii) *Symptoms (Figure 6.6)*
There is some evidence that *R. cerealis* and the closely related species *R. solani* cause pre-emergence damping off, especially if fields are waterlogged or particularly cold at the time of sowing. Some of the stem base browning, quite common on plants in the late

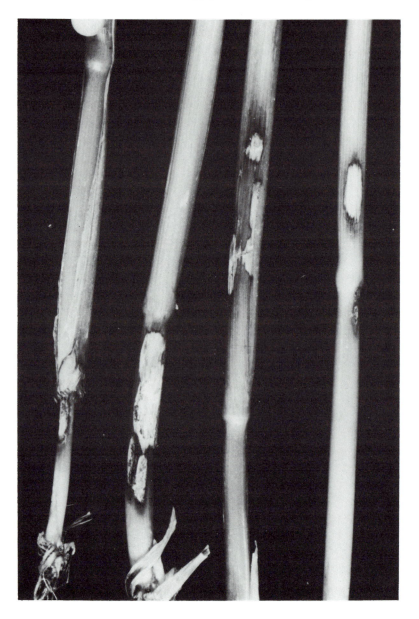

Figure 6.6. Sharp eyespot of cereals caused by *Rhizoctonia cerealis*. (A. Goulds, Rothamsted.)

autumn through to early spring, may be the result of sharp eyespot infection. However, *R. cerealis* tends to be less aggressive than eyespot and infection at this stage is likely to be superficial. The most characteristic symptoms of the disease in mature plants are the sharp eyespot lesions on the outer leaf sheaths. Many lesions may be found from the stem base to a height of about 30 cm. Lesions have a pale cream centre with a dark brown, sharply defined edge and there is no central black 'pupil' (cf. eyespot). Very often sharp eyespot lesions are quite superficial and do not penetrate the outer leaf sheath. However, particularly in recent years, the fungus has been observed on the stem itself where it can in certain circumstances cause structural damage in a similar way to the eyespot pathogen. There is some evidence to suggest that sharp eyespot is more common in crops where eyespot has been controlled. If sharp eyespot causes damage, both lodging and whiteheads may be a consequence. Occasionally white mycelium and sclerotia can be found in the infected stem cavity. Assessment of sharp eyespot may be undertaken using a key very similar to that used for eyespot (Clarkson & Cook, 1983).

(iv) *Disease cycle*

The fungus overwinters primarily as mycelium on infected stubble. Early-sown autumn crops, volunteers and some grass weeds may also act as sources of inoculum. Sclerotia in the soil have also been implicated as overwintering structures. The fungus does not produce spores readily and infection is initiated by the growth of hyphae from food sources onto cereal roots and stem bases. Infection may occur at any time during the growing season, but the disease is favoured by temperatures of around 9 °C. Acid, dry and sandy soils and early sowing also favour the disease. Sharp eyespot may be more severe in crops planted after potatoes. Cool autumn or spring temperatures may result in early infection by the fungus which can lead to severe sharp eyespot in mature plants. At the end of the season, infected stubble and trash will provide inoculum for successive cereal crops.

(v) *Significance*

Sharp eyespot has been observed on cereals at a low level for many years and was considered to be relatively insignificant in the UK. However, the disease began to become more widespread and severe at the start of the 1980s. The reasons for this are not clear. Perhaps modern varieties are more susceptible to sharp eyespot. Interestingly, the upsurge in sharp eyespot corresponded with an increase in

MBC usage for control of eyespot and there have been reports of MBCs encouraging sharp eyespot. Alternatively MBCs, by reducing eyespot, may simply have opened up a niche on the stem base which sharp eyespot filled. Sharp eyespot is widespread in the UK although it does not usually cause serious yield losses. Slight sharp eyespot is probably insignificant but moderate and severe sharp eyespot has been shown to reduce yield by 5 and 26%, respectively (Clarkson & Cook, 1983). Annual losses in the UK on average are probably less than 0.5%.

(vi) *Control*

Cultural control is probably the best way of reducing sharp eyespot. Rotation appears to have little effect on disease incidence, but it is probably unwise to follow a severely infected crop with another cereal. Late sowing may also reduce disease as would ploughing or burning infected stubble and debris.

Disease resistance in cereal species varies. The order of susceptibility has been demonstrated as rye, oats, wheat, barley. In addition, there is evidence that winter wheat varieties differ in their resistance to sharp eyespot. However, as with eyespot, the resistance is not particularly effective and the differences relate to severity of disease, not absolute resistance.

Chemical control of sharp eyespot is disappointing. Prochloraz (Sportak) may suppress the disease and there is also evidence for suppression by tolclofos-methyl (Rizolex). However, the degree of control is variable and is unlikely to be cost-effective.

Fusarium **foot rot**

(i) *Causal organism*
Fungi of genus *Fusarium*, primarily *F. nivale* (*Monographella nivalis*), *F. culmorum*, *F. avenaceum* (*Gibberella avenacea*) and *F. graminearum* (*G. zeae*).

(ii) *Host range*
Wheat, barley, oats, rye, grasses and numerous other plant species.

(iii) *Symptoms (Figure 6.7)*
There is some debate as to the precise symptoms caused by each of the different *Fusarium* species. However, all the species are

probably capable of causing pre- or post-emergence death of seedlings if soil or seeds are heavily contaminated. *Fusarium nivale* can cause snow mould in young plants. Plants die under long-lasting snow cover with mild soil temperatures, or when snow has melted. Brown smudge symptoms on cereal stem bases, which are frequently found between late autumn and spring, may be a result of *Fusarium* infection. As with sharp eyespot, *Fusarium* is less aggressive than eyespot and such lesions are likely to be confined to

Figure 6.7. *Fusarium* foot rot of cereals. Extreme right tiller shows severe brown foot rot symptoms. Brown basal nodes and dark brown vertical streaks on internodes visible on other tillers. (J. D. S. Clarkson, ADAS.)

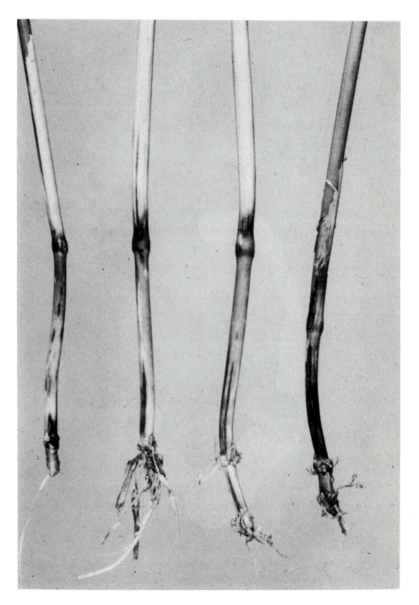

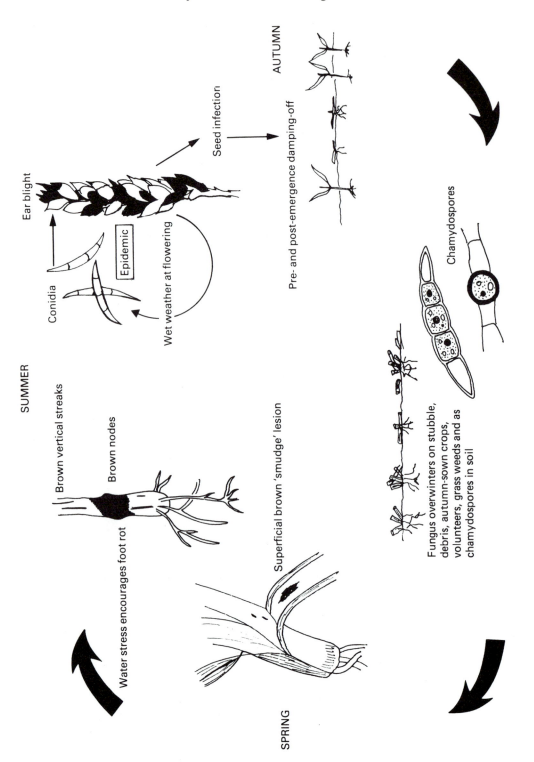

AUTUMN

Seed infection

Pre- and post-emergence damping-off

Ear blight

Conidia

Epidemic

Wet weather at flowering

WINTER

Chamydospores

Fungus overwinters on stubble, debris, autumn-sown crops, volunteers, grass weeds and as chamydospores in soil

SUMMER

Brown vertical streaks

Brown nodes

Water stress encourages foot rot

Superficial brown 'smudge' lesion

SPRING

the outer leaf sheath. On mature plants *Fusarium* infection may result in three fairly distinct symptoms. The first is brown foot rot, where the stem base becomes brown and water-soaked, which can result in lodging and whiteheads. This symptom is not common in the UK, although it can be quite common and damaging in the USA. More commonly in the UK, *Fusarium* causes browning on or around one of the basal nodes. Finally, dark brown vertical streaks on the basal internodes are also probably the result of *Fusarium* infection. If *Fusarium* colonisation is extensive, pink or orange spore masses may develop on the surface of the tissue.

Many assessment keys for *Fusarium* symptoms have been developed. The simplest method is to categorise tiller infection as slight, moderate or severe, using a key based on the eyespot or sharp eyespot assessment key.

(iv) *Disease cycle (Figure 6.8)*

There will be variations in aspects of the disease cycle presented depending on which *Fusarium* species is considered. Generally, however, all the cereal *Fusarium* species are ubiquitous in soil. Some can produce specially developed resting spores called chlamydospores. Most also have highly developed competitive saprophytic abilities which enable them rapidly to colonise debris and stubble. Hence *Fusarium* species mainly overwinter in the soil or on debris. Weeds, early-sown autumn crops, volunteers and infected seed may also act as overwintering sources of inoculum. From these bases, mycelium may grow directly into plant roots or stem bases, or spores may be produced which are splash-dispersed onto stem bases and the various symptoms then develop. Conditions which are conducive to severe disease development include drought stress. *Fusarium* species may infect the ear at the end of the growing season (see *Fusarium* ear blight), and they also remain in or rapidly colonise any stubble or debris remaining after harvest. Ploughing in of infected stubble will do little to reduce the disease.

(v) *Significance*

The disease is widespread in the UK and most cereal crops will have some *Fusarium* foot rot symptoms. However, disease severity in individual crops varies considerably and very few tillers show severe or potentially damaging disease symptoms. In the USA, the disease is much more of a problem and yield losses of up to 50% have been recorded in some cases, especially after drought stress.

Figure 6.8. Disease cycle of *Fusarium* foot rot and ear blight (*Fusarium* spp.).

(vi) *Control*

Cultural control offers very little. Rotation may help, but most of the cereal Fusaria will attack a wide range of other hosts. Burning infected stubble may reduce the amount of inoculum.

Chemical control may be effective at reducing early season infections arising from seed-borne inoculum. Some fungicides applied to control eyespot at GS 30–31 may also suppress *Fusarium* infection but there are no highly effective chemical control measures available. Products approved for use against *Fusarium* are listed below.

Active ingredient	*Product example*	
carboxin + imazalil + thiabendazole	Cerevax extra	
carboxin + thiabendazole	Cerevax	
ethirimol + flutriafol + thiabendazole	Ferrax	seed treatments
fuberidazole + imazalil + triadimenol	**Baytan IM**	
fuberidazole + triadimenol	Baytan	
maneb + zinc	Vassgro Manex	
prochloraz	Sportak	
carbendazim + prochloraz	Sportak Alpha	

Diagnosis of stem base diseases of cereals may be very difficult. Symptoms of eyespot, sharp eyespot and *Fusarium* during the early season are similar. It may be important to distinguish between the pathogens with regard to appropriate control measures. However, prior to GS 30 it is probably best to assume the worst and consider that any brown smudges are a result of the most aggressive pathogen of the three, *P. herpotrichoides* (eyespot). A suitable fungicide application may then follow at GS 30–33. In mature plants the situation may be less confusing, with typical lesions present. Quite often, however, stem bases become infected by a combination of pathogens, which can lead to confusion. As well as a keen eye and considerable experience, some laboratory isolation work involving examination of spores may be necessary to confirm the field diagnosis. Additionally, there may be some interaction between stem base pathogens, which can lead to atypical symptoms. In order to help in diagnosis of stem base pathogens Table 6.1 summarises their main diagnostic features.

Table 6.1. *Summary of main diagnostic features of stem base pathogens*

	Eyespot	Sharp eyespot	Fusarium foot rot
Symptoms on seedlings	Brown smudge on basal leaf sheath occasionally start of eyespot lesion. Sometimes black spot(s) on penetrated sheath	Superficial brown smudge (if present)	Superficial brown smudge (if present)
Symptoms on adult plants	Dark brown eye-shaped lesion, diffuse edge and central black 'pupil'. Usually below 1st node	Pale cream, sharply defined lesions. Usually above 1st node	Brown foot rot. Brown nodes. Dark brown vertical streaks
Mycelium in stem	Grey	White, if present. Sometimes sclerotia	Not usually inside stem. May be pink-orange spore masses on outside
Fungus on agar	Slow growing, grey colour	Fast growing. Hyaline (clear) mycelium, darkens with age	All fast growing. *F. nivale*: pale peach; *F. culmorum*: dark red; *F. avenaceum/F. graminearum*: yellow/orange/red
Spores	Long, thin, slightly curved (c. $40 \times 2\,\mu m$)	No spores, large mycelium branching at right angles. Slight constriction at branch	*F. nivale*: conidia sickle-shaped, 10–30×2.5–$5\,\mu m$ *F. culmorum*: conidia curved, 26–40×4–$6\mu m$ *F. avenaceum*: forms primary and secondary macrocondia, 10–70×3.5–$5\,\mu m$ *F. graminearum*: conidia quite rare in culture, 40–80×4.0–$6.5\,\mu m$

Powdery mildew

(i) *Causal organism*
Fungus, *Erysiphe graminis*.

(ii) *Host range*
E. graminis f.sp. *tritici* affects wheat;
E. graminis f.sp. *hordei* affects barley;
E. graminis f.sp. *avenae* affects oats;
E. graminis f.sp. *secalis* affects rye.
Note that within the formae speciales, there is further specialisation into races or pathotypes which can only attack particular varieties.

(iii) *Symptoms (Figure 6.9)*
Symptoms of powdery mildew can be found on all aerial parts of cereals, i.e. leaves, stems and ears, but leaves are most commonly infected. The first symptoms of infection, which can easily be overlooked, are chlorotic flecks on plant tissue. A white mildew pustule soon develops which quickly produces masses of spores and assumes a powdery appearance. If the plant tissue is tapped, clouds of mildew conidia, almost like dust, will be released. As the mildew pustules become older, they often change colour slightly and

Figure 6.9. Powdery mildew of cereals caused by *Erysiphe graminis*: close-up of pustules. (Schering Agriculture.)

acquire a grey or brown tinge. Towards the end of the season, dark coloured, sexually produced spore cases (cleistothecia) may be found embedded in the mildew pustules. Mildew on seedlings may be assessed using the following key:

Score	Resistance	Description
0	Very resistant	No sporulation
		No mycelium
1	Resistant	No sporulation
		Slight mycelium
2	Moderately resistant	Slight sporulation
		Moderate mycelium
3	Moderately susceptible	Moderate sporulation
		Abundant mycelium
4	Very susceptible	Abundant sporulation
		Abundant mycelium

This 'reaction-type' key is used in glasshouse tests to determine the genetic resistance of cereal varieties. Mildew on adult plants can be assessed by taking random samples from the field and estimating the percentage mildew infection on a suitable leaf by using a picture key (Chapter 3).

(iv) Disease cycle (Figure 6.10)

Mildew overwinters primarily as mycelium on green living material in the form of volunteers and autumn-sown crops. The cleistothecia produced during late summer are somewhat resistant to cold and drying out, and they may allow the fungus to survive for a time in the absence of a host. However, in temperate areas, fresh host plant material is nearly always available and cleistothecia are probably of secondary importance. In humid weather, cleistothecia release the sexually produced ascospores which can initiate infections on autumn-sown crops in the autumn and perhaps the spring. As temperatures rise in the spring, dormant mycelium starts to grow and conidia are quicky produced. Conidia germinate over a wide range of temperatures, from 5 to 30 °C, but a temperature of around 15 °C is optimal for germination together with a relative humidity above 95%. Free water inhibits spore germination. Under drier conditions, fresh conidia can be formed in about 7 days, and are wind-blown throughout the crop. Thus powdery mildew epidemics occur particularly during conditions of alternating wet and dry weather, with breezes to disperse spores. During a warm summer in the UK, when daytime temperatures exceed 25 °C, mildew growth is slowed down. The cleistothecia are formed in late summer. At the end of the season, volunteers and early autumn-sown crops may

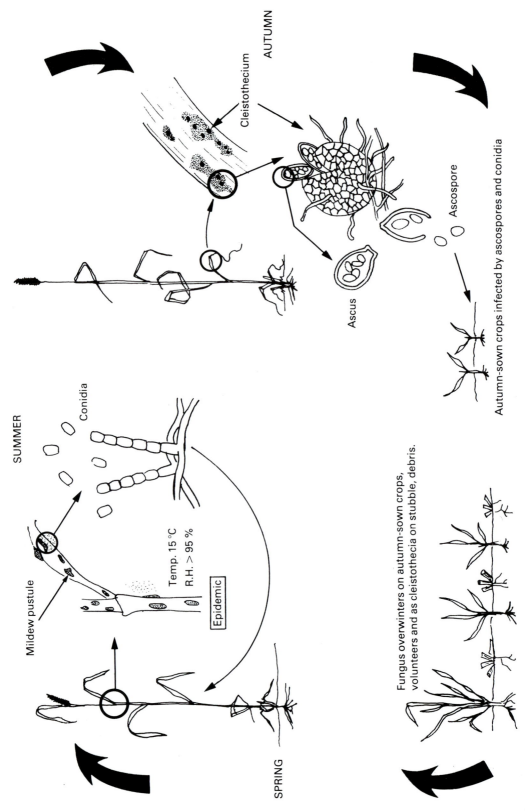

AUTUMN

Cleistothecium

Ascospore

Ascus

Autumn-sown crops infected by ascospores and conidia

WINTER

SUMMER

Conidia

Mildew pustule

Temp. 15 °C
R.H. > 95 %

Epidemic

Fungus overwinters on autumn-sown crops, volunteers and as cleistothecia on stubble, debris.

SPRING

become infected. Mildew is encouraged by very early sowing, particularly in barley. Badly mildewed plants in the autumn may be less resistant to winter frosts. Late-sown winter and spring crops may also be more prone to attack. Excessive nitrogen fertiliser also encourages the disease and mildew can be particularly severe in dense crops grown in sheltered humid places.

(v) *Significance*

Mildew is the most widespread leaf disease occurring on cereals in the UK. Barley tends to suffer more than wheat. Yield loss formulae suggest that for barley and oat powdery mildew the % yield loss $= 2.5 \sqrt{\%}$ mildew on leaf 3 at GS 58 and for wheat, % yield loss $= 2.0 \sqrt{\%}$ mildew. Yield losses of up to 25% in the USA and 20% in the UK have been reported in the field. Before the advent of highly effective systemic fungicides, powdery mildew was considered a superficial disease of little significance. Clearly this was not the case. Severe attacks of powdery mildew will reduce 1000-grain weight and may also reduce specific weight of grain. There is an increase in the significance of powdery mildew at present because of the development of fungicide-resistant strains.

(vi) *Control*

Cultural control of powdery mildew involves the eradication of volunteer cereals which act as overwintering sources of inoculum and disposal of stubble and debris which may be infested with cleistothecia. In addition, autumn- and spring-sown cereal crops should be isolated as much as possible from each other and excess nitrogen fertilizer avoided. Optimum amounts of other essential nutrients, particularly manganese, should be applied.

Disease resistance is very important. Wheat, barley and oat varieties exhibit a considerable range of resistance to the disease and, as mildew is widespread in the UK, it is wise to select a variety with a good disease resistance rating. If the variety chosen has poor mildew resistance because, for example, it has other desirable features, the crop should be monitored very frequently for mildew and it is likely that one or more applications of fungicide will be required.

In addition to selecting varieties with good mildew resistance, the farmer can reduce the spread of mildew from one field to adjacent fields by correct diversification of varieties. Good combinations of varieties are given in *Diversification Schemes* in the NIAB Farmers'

Figure 6.10. Disease cycle of powdery mildew (*Erysiphe graminis*).

Leaflet No. 8. Growing mixtures of varieties is also an option to reduce mildew.

Chemical control is widely practised. Systemic seed treatments which will protect against early attacks of powdery mildew can be used. This is advisable if a variety is particularly susceptible and sown early into light land. Winter barley in particular may need a fungicide application in the autumn if mildew is already spreading to young leaves. Heavy mildew infestations, particularly on light land can, in combination with severe winter frosts, significantly reduce plant populations. It is unlikely that autumn-sown wheat would require a fungicide in the autumn. In the spring, it may be wise to protect new growth with a fungicide, especially in crops which had severe mildew in the autumn. At any time during spring and summer before ear emergence, if mildew is found to be spreading to new tissue, a further fungicide application may be necessary. Fungicides approved for use against powdery mildew include the following:

Active ingredient	*Product example*
carbendazim + flutriafol	Early Impact
carbendazim + mancozeb	Kombat
carbendazim + maneb	Multi-W FL, Septal (wheat only)
carbendazim + maneb + sulphur	Bolda FL (wheat)
carbendazim + maneb + tridemorph	Cosmic FL
carbendazim + prochloraz	Sportak Alpha
carbendazim + propiconazole	Hispor 45 WP
chlorothalonil + fenpropimorph	Corbel CL
chlorothalonil + flutriafol	Impact Excel
copper oxychloride + maneb + sulphur	Ashlade SMC
ethirimol	Milstem (seed treatment, barley)
ethirimol + flutriafol + thiabendazole	Ferrax (seed treatment, barley)
fenpropidin	Patrol
fenpropimorph	Corbel, Mistral
fenpropimorph + iprodione	Sirocco
fenpropimorph + prochloraz	Sprint
ferbam + maneb + zineb	Trimanzone
fuberidazole + imazalil + triadimenol	Baytan IM (seed treatment, barley)

fuberidazole + triadimenol	Baytan (seed treatment)
mancozeb	Dithane 945
maneb + zinc	Chiltern Manex
manganese zinc ethylenebisdithiocarbamate complex	Trithac
nuarimol	Triminol
prochloraz	Sportak
propiconazole	Radar, Tilt
propiconazole + tridemorph	Tilt Turbo
pyrazophos	Missile
thiophanate-methyl	Cercobin Liquid (barley)
triadimefon	Bayleton
triadimenol	Bayfidan
triadimenol + tridemorph	Dorin
tridemorph	Bardew, Calixin, Ringer
triforine	Saprol (barley)

The efficacy and spectrum of activity of this very wide range of chemicals varies. As mildew frequently occurs with other diseases, both points are very important in selection. Reference should always be made to product manuals for further information. 'Suppression' is not as good as 'excellent control'.

One of the major reasons for differences in efficacy is the development of isolates of *E. graminis* which are resistant to fungicide groups. Resistance has been detected to the triazoles which include flutriafol, propiconazole, triadimefon and triadimenol. Hence these are often combined with another active ingredient, frequently a morpholine (e.g. tridemorph), in order to increase efficacy and reduce pressure on the pathogen to produce triazole-resistant isolates. There are also recommendations concerning the maximum numbers of triazole sprays in any season. It is wise not to use triazole sprays sequentially, and to diversify fungicide groups in a spray programme. Resistance to ethirimol was also detected in the early 1970s but recently there has been a shift back to sensitivity in the pathogen population (see Chemical control).

Yellow rust (stripe rust)

(i) *Causal organism*
Fungus, *Puccinia striiformis*.

(ii) Host range

P. striiformis f.sp. *tritici* on wheat;

P. striiformis f.sp. *hordei* on barley.

Yellow rust can also be found on many grass species. Within the formae speciales subspecies there are races or pathotypes which can only attack particular varieties.

(iii) Symptoms

Epidemics of yellow rust often start as disease foci in fields. Yellow/orange coloured pustules of yellow rust can occur on all aerial parts of the plant but are most frequently seen on leaves. Individual pustules (uredia) are roughly circular and about 0.5–1 mm in diameter. Their development may be preceded by an almost invisible chlorotic flecking. Aggregations of pustules tend to occur in stripes, between the veins of mature cereal plants; this effect is not apparent in softer young tissue. Pustules often give rise to chlorosis and later necrosis on leaves. During the summer it is possible for yellow rust infection of the ears to occur, resulting in the formation of masses of spores between the glume and the lemma. At the end of the season, black telia may form in patches of tissue killed by yellow rust uredia.

Yellow rust on seedlings may be assessed using the following key:

Score	Resistance	Description
0	Very resistant	Areas of chlorosis/ necrosis, no pustulation
1	Resistant	Very few pustules of low spore production with chlorosis/necrosis
2	Moderately resistant	Pustules of low spore production with or without chlorosis/necrosis
3	Moderately susceptible	Pustules of high spore production with chlorosis
4	Very susceptible	Pustules of high spore production without chlorosis

This 'reaction-type' key is used in glasshouse tests to determine the genetic resistance of cereal varieties. Yellow rust on adult plants can be assessed by taking random samples from the field and estimating the percentage yellow rust infection on a suitable leaf by using a picture key (Chapter 3).

(iv) Disease cycle (Figure 6.11)

Puccinia striiformis does not form any specialised resting spores to overwinter and requires living green plant material in order to survive. The fungus survives the winter mainly as dormant mycelium or uredia on volunteers and early autumn-sown crops. Should the host die, as a result of a hard frost, the fungus will also die. Yellow rust can survive freezing temperatures but it may be killed at temperatures below $-5\,°C$. In spring, particularly in cool moist weather, the fungus starts to grow again and more uredospores are produced. Temperatures of $10–15\,°C$ and a relative humidity of 100% are optimal for spore germination, penetration and production of new, wind-dispersed spores. This process may take as little as 7 days during such conditions. The disease cycle may therefore be repeated many times in one season. Yellow rust epidemics tend to start from disease foci in fields and, at least initially, the pattern of spread may be related to wind direction. As well as local inoculum initiating disease epidemics there is also a possibility of uredospores being blown into a crop from long distances. Yellow rust epidemics tend to occur in spring and early summer in the UK. The fungus is inhibited by temperatures over $20\,°C$. During late summer, telia may form which give rise to teliospores. The latter can form basidiospores but no alternate host has been found and the teliospores therefore have no function in the disease cycle.

(v) *Significance*

The disease is very sporadic in occurrence in the UK. It tends to occur most frequently in coastal areas which may have cool summer weather accompanied by regular mists. In isolated incidents where a susceptible variety has been grown, yield losses of up to 40% have been recorded. However, on average, yellow rust does not usually exceed 2% leaf area affected in the UK and yield losses are about 1%. A yield loss formula for yellow rust is % yield loss $= 0.4 \times$ % infection on flag leaf at GS 75 (King & Polley, 1976). Severe yellow rust can cause grain shrivelling and reduce specific weight.

(vi) *Control*

Cultural control methods are almost identical to those discussed for the control of powdery mildew (p. 183).

Disease resistance has, in the UK, been a very successful means of control, particularly in winter wheat. Severe epidemics of yellow rust were a frequent occurrence on winter wheat in the early 1970s because of a combination of very susceptible popular varieties and weather conducive to yellow rust epidemics. There was subsequently an increase in the effort to breed yellow rust resistant

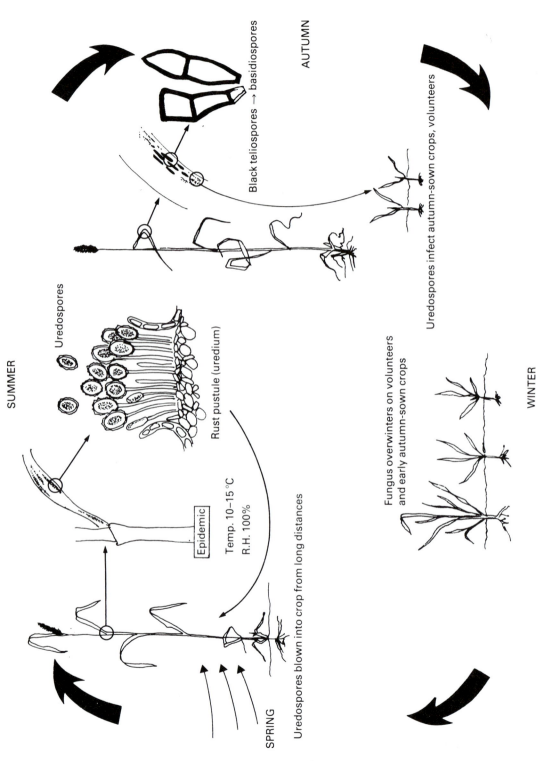

AUTUMN

Black teliospores → basidiospores

Uredospores infect autumn-sown crops, volunteers

SUMMER

Uredospores

Rust pustule (uredium)

Epidemic

Temp. 10–15 °C
R.H. 100%

Uredospores blown into crop from long distances

SPRING

Fungus overwinters on volunteers
and early autumn-sown crops

WINTER

varieties and many currently grown varieties have both major gene and polygenic (field) resistance. As with mildew, variety diversification and variety mixtures may also be used to good effect in yellow rust control.

Chemical control of yellow rust is likely to be routinely needed only on very susceptible varieties, when fungicides should be applied as soon as the disease appears. On less susceptible varieties, fungicides should be applied if weather conditions are favourable for disease development and there is evidence of the disease spreading in the crop. Many broad-spectrum fungicides applied for control of more common diseases such as powdery mildew will have activity against yellow rust. Fungicides approved for use against yellow rust include the following:

Active ingredient	*Product example*
benodanil	Calirus
carbendazim + flutriafol	Early Impact
carbendazim + mancozeb	Kombat
carbendazim + maneb	Delsene M Flowable, Multi-W FL, Septal
carbendazim + maneb + sulphur	Bolda FL (Wheat)
carbendazim + maneb + tridemorph	Cosmic FL
carbendazim + propiconazole	Hispor 45 WP
chlorothalonil + fenpropimorph	Corbel CL
chlorothalonil + flutriafol	Impact Excel
ethirimol + flutriafol + thiabendazole	Ferrax (seed treatment, barley)
fenpropidin	Patrol
fenpropimorph	Corbel, Mistral
fenpropimorph + iprodione	Sirocco
fenpropimorph + prochloraz	Sprint
fuberidazole + imazalil + triadimenol	Baytan IM (seed treatment, barley)
fuberidazole + triadimenol	Baytan (seed treatment)
maneb	Manzate
oxycarboxin	Plantvax 20 (Wheat)
propiconazole	Radar, Tilt
propiconazole + tridemorph	Tilt Turbo

Figure 6.11. Disease cycle of yellow rust (*Puccinia striiformis*).

Brown rust

(i) *Causal organism*
Fungi, *Puccinia recondita* on wheat and *P. hordei* on barley.

(ii) *Host range*
Wheat, barley, rye and grasses. *Puccinia hordei* has also been reported on *Ornithogalum, Leopoldia* and *Dipcadi* spp. *Ornithogalum* acts as an alternate host species in Israel, but is unimportant in the UK. Pathotypes of both *P. recondita* and *P. hordei* exist.

(iii) *Symptoms (Figure 6.12)*
Small orange/brown pustules may occur on all aerial parts of the plant, but are most frequently seen on the leaves. Individual

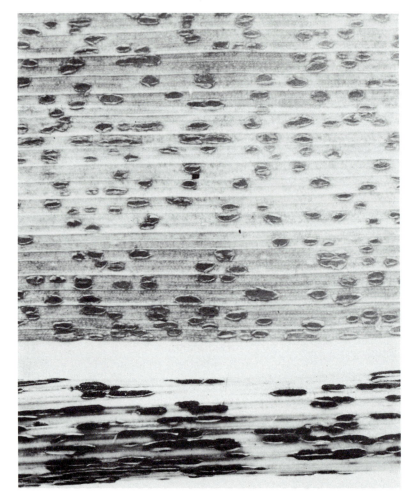

Figure 6.12. Brown rust of wheat caused by *Puccinia recondita* (top) and black stem rust caused by *Puccinia graminis* f.sp. *tritici* (bottom). (Illinois Agricultural Experiment Station.)

pustules may be slightly larger than those of yellow rust, at about 1×1–2 mm in diameter. Frequently a chlorotic halo surrounds individual pustules. In the early stages of the disease, pustules may be quite difficult to detect, but as the disease progresses leaves can develop an overall brown appearance. Pustules are scattered at random on leaves, not in stripes. Rings of secondary pustules surrounding a central primary infection site may be seen, and occasionally, when leaves begin to senesce, a 'green island' develops around individual pustules. Towards the end of the season, dark telia may develop. Assessment of brown rust can be undertaken as for yellow rust.

(iv) *Disease cycle*
The fungi overwinter primarily by the same means as yellow rust. During spring the disease may be slower to build up than yellow rust and a temperature of between 15 and 22 °C, accompanied by 100% relative humidity, is optimal for sporulation and spore germination. Therefore brown rust epidemics tend to occur during mid to late summer in the UK. Dry windy days which disperse spores and cool nights with dew favour the build-up of disease. Towards the end of the season, telia are formed and other spore stages have been found on a number of plants. These are considered insignificant in the UK although they may act as sources of new virulence.

(v) *Significance*
In the UK, brown rust of cereals is probably more common than yellow rust. National average levels of the disease rarely exceed 3% leaf area affected, but severe disease epidemics can occur in individual crops where yield losses of 30% have been reported. As brown rust is generally a 'higher temperature' disease than yellow rust, it usually occurs later in the season and is frequently found on the flag leaf of cereals. This increases its significance with regard to yield losses. A yield loss formula calculated for brown rust is very similar to that for yellow rust and the effects on grain quality would be similar.

(vi) *Control*

Cultural control methods and control by disease-resistant varieties operate in a similar way to those for yellow rust.

Chemical control of brown rust is also similar to that of yellow rust. However, as infection of the flag leaf is common, extra care must be taken to protect it from disease. Varieties which are highly susceptible to brown rust should receive a prophylactic spray as soon as the flag leaf has completely emerged and a second spray after ear emergence. Less susceptible varieties should be sprayed as soon as brown rust is seen spreading to the upper leaves.

Fungicides approved for the control of brown rust are largely the same as those approved for yellow rust.

Black stem rust

(i) *Causal organism*
Fungus, *Puccinia graminis*.

(ii) *Host range*
Puccinia graminis f.sp. *tritici* on wheat in the USA and Europe and barley in the USA. *Puccinia graminis* f.sp. *secalis* on barley in Europe.

Oats, rye and many grass species are also infected. The sexual phase of *P. graminis* is completed on its main alternate host *Berberis* spp. (barberry), although some *Mahonia* spp. may also act as alternate hosts. Races of *P. graminis* exist.

(iii) *Symptoms (Figure 6.12)*
Rust pustules (uredia) can occur on all aerial parts of the plant, but are most frequently seen on cereal stems. The pustules are larger than those of yellow and brown rust, about 2×5 mm, and are more elongated. They begin as a cinnamon brown colour and turn almost black as telia develop. A further characteristic symptom of black stem rust is a dry frayed epidermis surrounding the conspicuous pustule.

(iv) *Disease cycle*
Puccinia graminis overwinters either as telia or uredia. In cooler regions the fungus overwinters as telia on infested cereal debris. These spores germinate in the spring after being subjected to alternate freezing and thawing. The basidiospores, produced by teliospores, are dispersed short distances onto the alternate host, the barberry bush. Spermogonia then develop on barberry, which eventually release spermatia in a sticky insect-attractive fluid. Rainwater, dew and insects transmit the spermatia to sexually compatible receptive hyphae produced from heterothallic strains of the fungus. The fertilised hyphae then give rise to a further spore stage, the aeciospores, produced in bright orange aecia. Aecio-

spores are released in late spring and are wind-blown to the cereal host. The asexual spore stage (uredospore) develops on cereals giving rise to the characteristic black stem rust pustules which are uredia. Uredospores may be wind-blown over long distances, resulting in infections many miles from the source of inoculum. The uredospore stage may survive all year round on its cereal host in warmer climates. Black stem rust is a disease of warmer climates where it occurs late in the season. Temperatures above 20 °C together with 100% relative humidity are optimal for disease development. The pathogen is inhibited by temperatures below 15 °C.

(v) *Significance*

Where it occurs (North America, India, Australia and parts of North Africa and the Mediterranean basin) black stem rust is probably the most serious of the rust diseases. It is of little significance in the UK. Ruptures of the epidermis can result in tissue drying out, assimilates not reaching the ear and whiteheads developing. Numerous stem lesions can also result in lodging. Long-term average yield losses in some areas of Europe have been reported at 10% but yield losses in individual crops may be much higher.

(vi) *Control*

Cultural control methods which include eradication of the alternate host, barberry, have been widely practised over many years in order to reduce black stem rust epidemics (see Cultural control, p. 95).

Disease resistance is a topic of much interest, particularly in the USA, where much effort has been directed at producing black stem rust-resistant cereal varieties. Unfortunately, in many cases resistance has been eroded after only a few years by the development of new races of *P. graminis* which could attack the new varieties. Effort is now directed at producing varieties with polygenic (durable) resistance.

Chemical control methods have not been widely developed, mainly because the disease is a problem primarily in areas where less intensive agriculture is practised and fungicide applications may not be economical. Many of the fungicides effective against yellow and brown rust would also have activity against black stem rust.

Crown rust

(i) *Causal organism*
Fungus, *Puccinia coronata* var. *avenae*.

(ii) *Host range*
Oats and other grass species. *Rhamnus catharticus* and *Frangula alnus* may act as alternate hosts.
    Races of *P. coronata* var. *avenae* exist.

(iii) *Symptoms*
Bright orange elongated rust pustules 1–5 mm in length occur on all aerial parts, but primarily on the leaves of oats. Pustules often occur in patches, forming dense orange areas on the leaf. The disease is generally seen on adult plants, and towards the end of the season black telia may also be found, particularly on leaf sheaths.

(iv) *Disease cycle*
The pathogen overwinters as uredia, telia and dormant mycelium on host plant debris, volunteers and autumn-sown crops. Spring infections are initiated mainly from these overwintering sources but aeciospores produced on the alternate host *Rhamnus* may also initiate the disease, especially in parts of Eastern Europe. During the spring, germinating crown-shaped teliospores can give rise to basidiospores which infect the alternate host. These can produce aeciospores which re-infect oats in late spring to early summer. Disease epidemics progress most quickly at temperatures around 25 °C when free water is available.

(v) *Significance*
Crown rust is probably the most important disease of oats world-wide. Losses of up to 20% have been reported in some areas of the USA. In the UK, only sporadic outbreaks of crown rust occur, most often in the south-west.

(vi) *Control*
Cultural control methods, including destruction of *Rhamnus* bushes near to oats, and elimination of volunteers, stubble and debris will all help to reduce disease.

Disease resistance has been incorporated into oat varieties, but again there have been problems with resistance breakdown. More durable or 'slow-rusting' varieties are being actively sought.

Chemical control is probably unjustified on economic grounds except in high value seed crops. Many of the broad-spectrum systemic fungicides used to control powdery mildew on oats will also be effective against crown rust.

Leaf and glume blotch (*Septoria nodorum*)

(i) *Causal organism*
Fungus, *Septoria nodorum* (*Leptosphaeria nodorum*).

(ii) *Host range*
Mainly wheat, but also barley and rye.

(iii) *Symptoms (Figure 6.13)*
Septoria nodorum can infect seedlings, resulting in water-soaked, dark green areas later becoming necrotic. Twisted, distorted and stunted plants, with occasional necrotic protruberances on the hypocotyl, may also occur. On mature leaf tissue the first symptoms of infection are small necrotic lesions. Later these develop into brown oval lesions surrounded by a chlorotic halo. These lesions frequently coalesce to produce large areas of dead, dry and sometimes split tissue. Spore cases (pycnidia) form within affected

Figure 6.13. Leaf blotch of wheat caused by *Septoria nodorum*. (J. D. S. Clarkson, ADAS.)

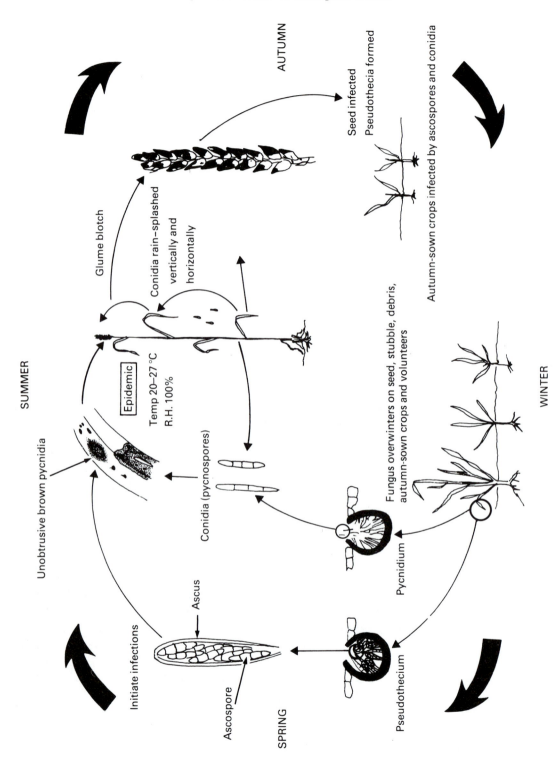

AUTUMN

Seed infected
Pseudothecia formed

Autumn-sown crops infected by ascospores and conidia

Glume blotch

Conidia rain–splashed
vertically and
horizontally

SUMMER

Epidemic
Temp 20–27 °C
R.H. 100%

Conidia (pycnospores)

Fungus overwinters on seed, stubble, debris,
autumn-sown crops and volunteers

WINTER

Unobtrusive brown pycnidia

Pycnidium

Ascus

Initiate infections

Ascospore

SPRING

Pseudothecium

tissue, but these are brown, unobtrusive and very difficult to see without magnification. During periods of wet weather, it may be possible to see salmon-pink gelatinous spore masses, called cirri, exuding from the pycnidia. *Septoria nodorum* can also infect the ears, particularly of wheat, causing glume blotch. Dark brown patches like burn-marks develop on the glumes, which later become purple-brown. Glume blotch symptoms are easiest to see on green ears. Pycnidia formed on the ear may be a little easier to see than those on the leaves.

Picture keys published by ADAS are available for assessing both *Septoria* leaf and glume blotch.

(iv) *Disease cycle (Figure 6.14)*

Septoria nodorum does not require green living plant material on which to overwinter. It survives as dormant mycelium, pycnidia and pseudothecia on seed, stubble, debris, autumn-sown crops and volunteers. In the absence of crop debris, initial infections in the autumn or spring may result from wind-borne ascospores released from pseudothecia long distances away. As temperatures rise and humidity increases, asexual pycnospores are produced from pycnidia. These are splash-dispersed vertically up the infected plant and horizontally from plant to plant. Temperatures of 20–27 °C, together with 100% relative humidity, are optimal for spore production and germination; periods of rain are essential for spore dispersal. The disease cycle can be completed in 10–14 days during such conditions. Spores produced from pseudothecia and pycnidia, which develop on the flag leaf and ear at the end of the season, can initiate infection in early autumn-sown crops and volunteers and may also remain dormant for the winter. Glume blotch symptoms on the ear can lead to seed infection.

(v) *Significance*

Septoria nodorum is one of the most serious pathogens on cereals in the UK; it is a widespread disease in most years. Further, as a late-season disease, its effect on yield can be quite dramatic. Yield losses in excess of 50% have been reported as a result of severe attacks of *S. nodorum*, although average annual losses in the UK probably do not exceed 10%. *Septoria nodorum*, particularly on wheat ears, reduces both 1000-grain weight and specific weight, but may improve protein content of the grain.

Figure 6.14. Disease cycle of leaf and glume blotch (*Septoria nodorum* = *Leptosphaeria nodorum*).

(vi) *Control*

Cultural control of *S. nodorum* includes correct disposal of crop debris by burning or ploughing.

Disease resistance in winter wheat varieties is disappointing, with only tolerance or a moderate degree of resistance exhibited. A few varieties are particularly susceptible.

Chemical control of *S. nodorum* may be essential. Initially, seed infection can be reduced by a number of seed treatments. Decisions to apply fungicides during the growing season are based on observations of symptoms spreading to younger tissue or weather conditions (see later). It is essential to protect the flag leaf and the ear from disease. Fungicides approved for use against *Septoria* are included in the list below. There is no distinction between efficacy against *S. nodorum* and against *S. tritici*, therefore product manuals should be checked for any specific recommendations.

Active ingredient	*Product example*
carbendazim + chlorothalonil	Bravocarb
carbendazim + flutriafol	Early Impact
carbendazim + mancozeb	Kombat
carbendazim + maneb	Delsene M Flowable, Multi-W FL
carbendazim + maneb + sulphur	Bolda FL
carbendazim+maneb+ tridemorph	Cosmic FL
carbendazim + prochloraz	Sportak Alpha
carbendazim + propiconazole	Hispor 45 WP
carboxin + thiabendazole	Cerevax (seed treatment)
chlorothalonil	Bravo 500
chlorothalonil + fenpropimorph	Corbel CL
chlorothalonil + flutriafol	Impact Excel
copper hydroxide	Chiltern Kocide 101
copper oxychloride + maneb + sulphur	Ashlade SMC
fenpropimorph + iprodione	Sirocco
fenpropimorph + prochloraz	Sprint
ferbam + maneb + zineb	Trimanzone
fuberidazole + triadimenol	Baytan (seed treatment)
iprodione	Rovral Flo
iprodione + thiophanate-methyl	Compass

mancozeb	Dithane 945
maneb	Manzate
maneb + zinc	Manex
manganese zinc ethylenebisdithiocarbamate complex	Trithac
nuarimol	Triminol
prochloraz	Sportak
propiconazole	Radar, Tilt
propiconazole + tridemorph	Tilt Turbo

Forecasting disease may result in the most economical and effective use of fungicides. Spread of the disease depends on rainfall. Hence a high risk period for *Septoria* has been defined as three consecutive rain days in which at least 10 mm have fallen in the two weeks previous to flag leaf emergence (GS 37). A fungicide should then be applied as soon as possible. Applications of fungicide may be required before GS 37 if the disease is spreading.

Leaf blotch (*Septoria tritici*)

(i) *Causal organism*
Fungus, *Septoria tritici* (*Mycosphaerella graminicola*).

(ii) *Host range*
Mainly wheat, but also occasionally on rye and some grass species.

(iii) *Symptoms (Figure 6.15)*
Symptoms of *S. tritici* can be seen very early in the growing season in most years. On young autumn-sown wheat, water-soaked patches which quickly turn brown and necrotic may be evident before and after winter on the lowest leaves. The most characteristic feature of *S. tritici* is the development of visible black pycnidia in the brown lesions. Pycnidia are particularly common on dead overwintering leaves of winter wheat. During spring, pycnidia may ooze spores in a gelatinous creamy-white matrix, especially during humid weather. Lesions on mature plants are brown and frequently surrounded by a chlorotic halo. Lesions sometimes appear to be restricted by veins and this often gives them an elongated rectangular appearance. The presence of black pycnidia in lesions is indicative of *S. tritici*. Lesions may coalesce leading to large areas of brown tissue. During severe disease epidemics, it is possible for infection of the ears to occur. Bleached necrotic lesions on the

Figure 6.15. Leaf blotch
of wheat caused by
Septoria tritici. Note
presence of black
pycnidia. (Illinois
Agricultural Experiment
Station.)

glumes accompanied by many pycnidia are the most typical symptoms.

Levels of *S. tritici* are assessed in the same way as *S. nodorum*.

(iv) *Disease cycle*

The disease cycle of *S. tritici* is very similar to that of *S. nodorum*, although *S. tritici* can go through its disease cycle at slightly lower temperatures (15–20 °C optimum), and requires longer periods of high humidity to initiate infection. The disease may be evident earlier in the season than *S. nodorum*.

(v) *Significance*

Since 1985 there have been some particularly severe attacks of *S. tritici* on winter wheat in the UK and losses of 40% have been reported in severely affected crops. However, average losses in the UK over the last decade were much lower because of the sporadic occurrence of disease epidemics. The percentage yield loss for both *Septoria* diseases has been estimated to equal the percentage disease on the flag leaf at GS 75. The effects of *S. tritici* on quality of cereal grain are similar to those recorded for *S. nodorum*.

(vi) *Control*

Methods of control are similar to those used for *S. nodorum*. There are, however, a few notable differences. Firstly, some varieties of winter wheat differ in their resistance to *S. nodorum* and *S. tritici*. Secondly, there are some problems with fungicide-resistant isolates of *S. tritici*. After the widespread use of fungicides, many MBC-based, during attempts to halt a severe disease epidemic during 1985 in the UK, MBC resistance in *S. tritici* was detected. It is now considered to be widespread and fungicides containing MBC alone or in mixtures are no longer recommended for control of *S. tritici*. Fungicide applications for *S. tritici* are probably most beneficial at GS 33–45.

It may be important to discriminate between the two diseases. A summary of the main differences is given in Table 6.2. Diagnosis is best confirmed by microscopic examination of spores: there are marked differences in spore size between the two species.

Table 6.2. *Summary of main differences between* Septoria nodorum *and* Septoria tritici

	Septoria nodorum	*Septoria tritici*
Symptoms	Brown oval lesions with chlorotic margin	Brown lesions in stripes, chlorotic margin
Pycnidia	Small, brown, unobtrusive	Large, black, prominent
Spore masses (cirri)	Salmon-pink	Creamy white
Pycnospores (conidia)	Slightly curved, short and *c.* $19 \times 4\,\mu m$	Slightly curved, long and thin $43–70 \times 1.5–2\,\mu m$
Ascospores	$24–32 \times 4–6\,\mu m$	$10–15 \times 2.5–3\,\mu m$
Fungus on agar	Fast growing, brown cultures with even edge	Very slow growing, dark colony with uneven edge

Leaf blotch or scald (*Rhynchosporium*)

(i) *Causal organism*

Fungus, *Rhynchosporium secalis*

(ii) *Host range*

Mainly barley, although rye and many grass species can also be affected. There is some evidence for the existence of races of R. *secalis* on barley.

(iii) *Symptoms (Figure 6.16)*

The first sign of *Rhynchosporium* infection is a small (*c.* 1 cm) pale grey-green, water-soaked patch on the leaves. The lesion enlarges, its centre becomes pale grey-brown and a dark brown wavy border develops. Lesions may coalesce giving rise to large areas of necrotic tissue. The dark borders of individual lesions may still be visible within the dead tissue. Most *Rhynchosporium* lesions develop on leaves although leaf sheaths and glumes may also be affected. It is common to find lesions at the junction of the leaf and the stem (leaf axil). Such lesions can be particularly damaging, as they can cause the leaf to lose its natural erect position thereby reducing photosynthesis, or they may kill the leaf.

(iv) *Disease cycle (Figure 6.17)*

Figure 6.16. Leaf blotch or scald of barley caused by *Rhynchosporium secalis*. Note development of lesions in the leaf axil.

The fungus overwinters as mycelium on stubble, debris, seed, autumn-sown crops and volunteers. On stubble, *Rhynchosporium* may survive for up to 12 months. Primary infections in early spring arise mainly from splash-dispersed spores produced on affected

crop debris. The optimum temperature for spore production ranges from 10 to 20 °C. Free moisture, or relative humidities above 95%, are also necessary; infection requires the presence of free water and is highly likely if the leaf is wet for more than 24 hours. The disease cycle will take approximately 14 days during optimum weather conditions. If temperatures exceed 24 °C and there is a period of dry weather, the disease will develop no further. *Rhynchosporium* spores may be rain-splashed to the ear, where seed infection is possible. At the end of the season, the fungus persists on stubble and debris, but does not compete very well as a saprophyte in the soil. Early-sown autumn crops may become infected by spores produced on debris during mild wet autumn and winter weather.

An ADAS picture key is available for assessment of *Rhynchosporium*.

(v) *Significance*

Disease epidemics of *Rhynchosporium* occur sporadically in the UK. Annual national yield losses on average probably do not exceed 1%. However, in individual crops yield reductions of between 30 and 40% have been reported from many parts of the world. Yield loss formulae suggest that the percentage yield loss is equal to half the percentage infection on leaf 2 at GS 75. There is some evidence to suggest that *Rhynchosporium*, in combination with other foliar diseases of barley including net blotch and mildew, reduces 1000-grain weight and specific weight.

(vi) *Control*

Cultural control involves correct disposal of crop debris, by burning or ploughing, and elimination of volunteers. Crop rotation may also reduce disease levels. Optimum, not excess, rates of nitrogen and optimum time of sowing are important.

Disease resistance is available in both winter and spring varieties of barley. The resistance is mainly polygenic, although less durable major gene resistance has been identified. The few very susceptible varieties should be avoided in regions where *Rhynchosporium* epidemics are common.

Chemical control is widely practised in the UK. In winter barley, a systemic seed treatment may be used both to reduce seed infection and protect against early attacks of *Rhynchosporium*. However, a spray in the autumn should be applied if *Rhynchosporium* reaches

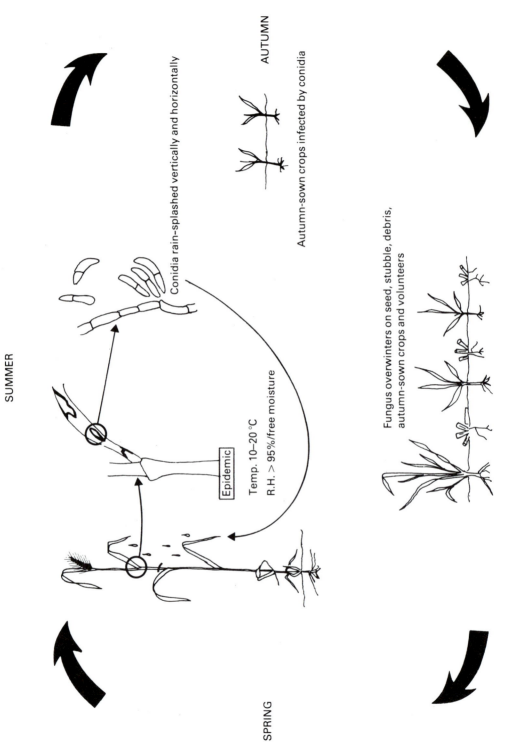

SUMMER

AUTUMN

WINTER

SPRING

Conidia rain-splashed vertically and horizontally

Autumn-sown crops infected by conidia

Epidemic

Temp. 10–20 °C

R.H. > 95%/free moisture

Fungus overwinters on seed, stubble, debris, autumn-sown crops and volunteers

10% leaf area affected on lower leaves. During spring and summer the crop should be examined carefully and, if *Rhynchosporium* lesions start to appear on new growth, a fungicide spray should be considered. It is important to protect the flag leaf from infection. Fungicides approved for the control of *Rhynchosporium* include the following:

Active ingredient	*Product example*
benomyl	Benlate
carbendazim	Bavistin, Delsene 50DF, Derosal Liquid
carbendazim + chlorothalonil	Bravocarb
carbendazim + flutriafol	Early Impact
carbendazim + mancozeb	Kombat
carbendazim + maneb	Multi-W FL
carbendazim + maneb + sulphur	Bolda FL
carbendazim + maneb + tridemorph	Cosmic FL
carbendazim + prochloraz	Sportak Alpha
carbendazim + propiconazole	Hispor 45 WP
chlorothalonil	Bravo 500, Bombardier
copper oxychloride + maneb + sulphur	Ashlade SMC
ethirimol + flutriafol + thiabendazole	Ferrax (seed treatment)
fenpropidin	Patrol
fenpropimorph	Corbel, Mistral
fenpropimorph + iprodione	Sirocco
fenpropimorph + prochloraz	Sprint
ferbam + maneb + zineb	Trimanzone
fuberidazole + imazalil + triadimenol	Baytan IM (seed treatment)
fuberidazole + triadimenol	Baytan (seed treatment)
iprodione + thiophanate-methyl	Compass
maneb	Manzate
maneb + zinc	Vassgro Manex
nuarimol	Triminol
prochloraz	Sportak
propiconazole	Radar, Tilt
propiconazole + tridemorph	Tilt Turbo
pyrazophos	Missile
thiophanate-methyl	Cercobin Liquid

Figure 6.17. Disease cycle of leaf blotch (scald) (*Rhynchosporium secalis*).

triadimefon   Bayleton
triadimenol   Bayfidan
triadimenol + tridemorph   Dorin

Net blotch

(i) *Causal organism*
Fungus, *Pyrenophora teres* (*Drechslera teres*).

(ii) *Host range*
Barley.

(iii) *Symptoms (Figure 6.18)*
Net blotch is most frequently seen on leaves of young autumn-sown
barley and volunteers during late autumn to early spring. Initially,
lesions appear as minute spots or streaks. These spread to form
narrow dark brown longitudinal streaks. Transverse streaks, giving
lesions a net-like appearance, may also be formed. Lesions are
often surrounded by areas of chlorosis and large areas of dead tissue
may occur. Another form of net blotch, a spot form, was discovered
in Denmark in 1971 and has also been seen in a few crops in the UK.
Symptoms of the spot form consist of dark brown elliptical lesions,
again surrounded by a chlorotic margin.
    An ADAS picture key is available for assessment of net blotch.

(iv) *Disease cycle*
The primary source of overwintering inoculum is infected stubble
and crop debris. However, the fungus can also overwinter on seed,
volunteers and autumn-sown crops as resting mycelium and
pseudothecia. Most infections occur at relatively low temperatures
(below 15 °C) and the fungus is inhibited by temperatures over
20 °C. Between 10 and 30 hours of high humidity are also required
for sporulation and infection. The large conidia of *P. teres* are
dispersed mainly by wind although some water dispersal may also
occur. Primary infections may also arise from wind-dispersed
ascospores released from pseudothecia. The disease cycle of net
blotch may be completed in 10–14 days during optimal weather
conditions. In dry summer weather, however, disease development
is inhibited.

(v) *Significance*
In the UK, severe epidemics of the disease are uncommon. In the
early 1980s, particularly 1981, net blotch was a very significant

disease and epidemics occurred throughout the growing season resulting in yield losses of between 10 and 40%. Nationally, annual losses now rarely exceed 1% on average. For the effects of net blotch on quality of barley grain, see *Rhynchosporium*.

(vi) *Control*

Cultural control methods are similar to those used for control of *Rhynchosporium*.

Disease resistance is available, particularly in varieties of winter barley. Most currently grown UK varieties have good resistance to net blotch.

Figure 6.18. Net blotch of barley caused by *Pyrenophora teres*.

Chemical control may initially be effected by systemic seed treatments. Net blotch may require treatment in the autumn if 20% area of lower leaves is affected by the disease. Crops should also be monitored for the disease in the spring. If net blotch can easily be found in the crop, a fungicide application should be considered. The optimum time for fungicide application is at GS 37. It is unlikely that fungicides will be required during late summer, but crops should be monitored. Fungicides approved for the control of net blotch include the following:

Active ingredient	*Product example*
2-methoxyethylmercury acetate	Panogen M (seed treatment)
carbendazim + flutriafol	Early Impact
carbendazim + mancozeb	Kombat
carbendazim + maneb	Multi-W FL
carbendazim + maneb + tridemorph	Cosmic FL
carbendazim + prochloraz	Sportak Alpha
carbendazim + propiconazole	Hispor 45 WP
carboxin + imazalil + thiabendazole	Cerevax Extra (seed treatment)
carboxin + phenylmercury acetate	Murganic RPB (seed treatment)
ethirimol + flutriafol + thiabendazole	Ferrax (seed treatment)
fenpropimorph + iprodione	Sirocco
fenpropimorph + prochloraz	Sprint
ferbam + maneb + zineb	Trimanzone
fuberidazole + imazalil + triadimenol	Baytan IM (seed treatment)
gamma HCH + phenylmercury acetate	Mergamma 30 (seed treatment)
iprodione	Rovral Flo
iprodione + thiophanate-methyl	Compass
maneb	Maneb, Manzate
nuarimol	Triminol
phenylmercury acetate	Ceresol (seed treatment)
prochloraz	Sportak
propiconazole	Radar, Tilt
propiconazole + tridemorph	Tilt Turbo
pyrazophos	Missile
triforine	Saprol

Barley yellow dwarf virus (BYDV)

(i) *Causal organism*
Virus, BYDV whose vectors in the UK are either the bird cherry aphid (*Rhopalosiphum padi*) or the grain aphid (*Sitobion avenae*).

(ii) *Host range*
Wheat, barley, oats and many grass species.

(iii) *Symptoms*
BYDV symptoms differ depending on the crop, but a general dwarfing occurs in all species. In wheat, leaves turn yellow and then red. This colour change usually starts at the tip of the leaf and progresses backwards. In barley a bright golden yellowing, again starting at the leaf tip, predominates. Apparently healthy green stripes of tissue may also be seen. In oats, a distinctive red-purple discoloration of the leaves occurs. Affected plants may be invaded by sooty moulds before harvest. Random plants may be affected in fields or patches of affected plants develop.

(iv) *Disease cycle (Figure 6.19)*
During winter, BYDV persists in early-sown autumn cereals, volunteers and its aphid vector. The aphid vector overwinters primarily on the same hosts. Spread of the virus is entirely dependent on the behaviour of its vector. In the spring, aphids migrate from their overwintering host grasses into nearby cereal crops to feed where infected (viruliferous) aphids transmit the virus. The aphids can acquire the virus in as little as 30 minutes, but generally around 24 hours' feeding time is required. BYDV does not multiply in the vector but a latent period of 1–4 days is required before it can be transmitted. Aphids may remain infective for several weeks. Symptoms of BYDV develop approximately 2 weeks after infection. Moist, relatively cool conditions (10–18 °C) favour aphid multiplication, secondary infection and migration. During late summer, aphids may migrate back to grass or colonise early autumn-sown crops. Migration usually stops in late October.

(v) *Significance*
Early widespread infections can reduce yields by as much as 40%. Yields of individual affected plants may be negligible. BYDV is considered to be the most important and widespread virus disease of barley.

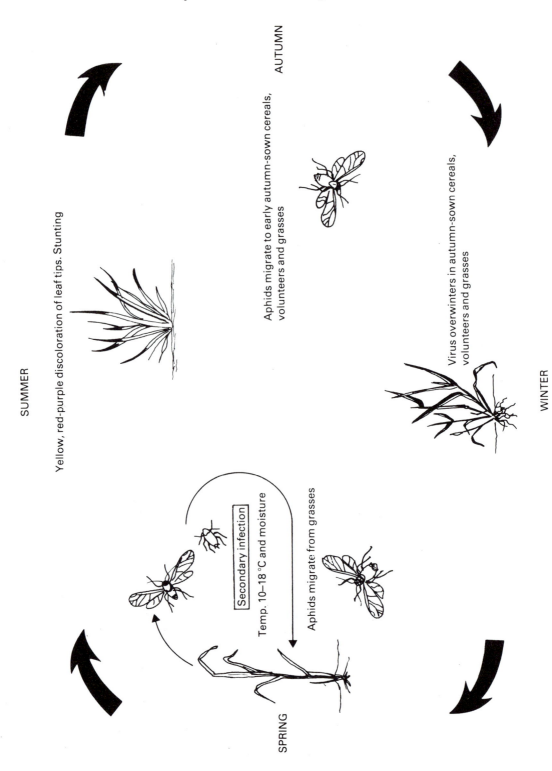

AUTUMN

Aphids migrate to early autumn-sown cereals, volunteers and grasses

Virus overwinters in autumn-sown cereals, volunteers and grasses

WINTER

SUMMER

Yellow, red-purple discoloration of leaf tips. Stunting

Secondary infection

Temp. 10–18 °C and moisture

Aphids migrate from grasses

SPRING

(vi) *Control*

Cultural control, aimed at eliminating reservoirs of infection such as grasses, is difficult in practice. Infected volunteers and grass weeds can, however, easily be eliminated before crop emergence. Most severe attacks of BYDV occur on very early-sown crops, therefore such crops carry a high risk of infection and should be protected (chemical control). Cereals which follow permanent pasture or grass leys are also high risk crops. This can be reduced somewhat by chemically killing the grass 7–10 days before cultivation and allowing 2 weeks before sowing cereals.

Disease resistance has recently been introduced into winter barley varieties which are included on the NIAB *Recommended Varieties of Cereals* (Farmers' Leaflet No. 8).

Chemical control is aimed at eliminating the aphid vector before it feeds on and infects the cereal host. Crops 'at risk' from BYDV should be treated with an appropriate aphicide in late October–early November. Checking crops for aphids is worthwhile as is taking heed of ADAS warnings regarding BYDV infection (forecasting). Aphicides approved for control of BYDV vectors include the following:

Active ingredient	*Product example*
deltamethrin	Decis
demeton-S-methyl	Metasystox 55
cypermethrin	Ambush C

Forecasting disease outbreaks of BYDV is undertaken by monitoring the populations of migrating viruliferous aphids. Local warnings are then issued by ADAS on the basis of these data.

Barley yellow mosaic virus (BaYMV)

(i) *Causal organism*
Virus, BaYMV whose vector is a soil-borne fungus, *Polymyxa graminis*.

Figure 6.19. Disease cycle of barley yellow dwarf virus (BYDV). Vector: aphid species.

(ii) *Host range*
Winter barley only.

(iii) *Symptoms (Figure 6.20)*

First symptoms of BaYMV can be seen from late December to March. Chlorotic streaks appear in both old and new leaves. These streaks, which are quite difficult to detect, may become more obvious if they become necrotic. Plants frequently have a spiky appearance as the leaf edges roll inwards. On a field scale, the disease can be seen as patches of chlorotic plants. It is common for

Figure 6.20. Barley yellow mosaic of winter barley caused by barley yellow mosaic virus. Note chlorotic streaks in old and young leaf. Also note leaf edges rolling inwards.

many of these early symptoms to disappear during summer, but affected mature plants are often stunted and poor-yielding.

(iv) *Disease cycle (Figure 6.21)*

The disease cycle of the virus is essentially the disease cycle of the fungal vector. *Polymyxa graminis* can survive in soil for many years as resting sporangia. The virus can also remain dormant in the fungal resting spore. Biflagellate zoospores are released from resting spores which swim in soil moisture and infect barley roots via root hairs or epidermal cells. Virus-contaminated zoospores will transmit the virus to the barley. Plasmodia are formed in affected barley roots and more zoospores may be released, resulting in secondary infections of nearby plants. Towards the end of the season the plasmodia develop into grape-like clusters of resting spores which are released into the soil, or persist on stubble and crop debris. Contaminated soil can be spread by cultivations, resulting in patches of affected plants following cultivation lines. Contaminated soil can also be spread from field to field on machinery and boots.

(v) *Significance*

BaYMV was first identified in Japan in 1940. It was reported in West Germany in 1978 and in France and the UK in 1980. It has now spread quite extensively through the UK and poses a major threat to winter barley production once it becomes established in a field. Yield losses of between 10 and 90% have been reported depending upon variety, climate, soil type and level of soil infestation.

(vi) *Control*

Cultural control of BaYMV is difficult. The longevity of the fungal resting spores makes crop rotation an ineffective means of control. Obviously care should be taken with regard to movement of contaminated soil, and machinery should be cleaned after use on a contaminated field.

Chemical control is not feasible.

Disease resistance. The only method of control is by using BaYMV-resistant varieties. A few of them have become available in the UK and some are included on the NIAB list of winter barley varieties. The durability of resistance to BaYMV has recently come

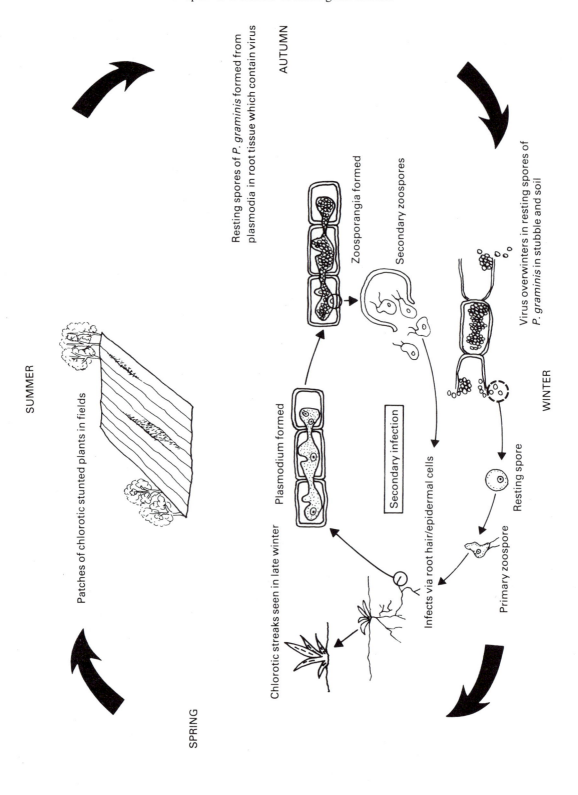

AUTUMN

Resting spores of *P. graminis* formed from plasmodia in root tissue which contain virus

Zoosporangia formed

Secondary zoospores

Virus overwinters in resting spores of *P. graminis* in stubble and soil

SUMMER

Plasmodium formed

Secondary infection

WINTER

Patches of chlorotic stunted plants in fields

Resting spore

Chlorotic streaks seen in late winter

Infects via root hair/epidermal cells

Primary zoospore

SPRING

into question, however, with one report of field infection of a resistant variety.

Loose smut

(i) *Causal organism*
Fungi, *Ustilago nuda* attacks wheat and barley (sometimes referred to as *U. tritici* on wheat); *U. avenae* attacks oats.

(ii) *Host range*
Wheat, barley, oats.

(iii) *Symptoms (Figure 6.22)*
The most obvious symptoms of loose smut occur when plants have reached the stage of ear emergence. Quite often smutted heads emerge a little before healthy ones and they may also be slightly taller. Normally the entire ear is replaced by a mass of black sooty smut spores. Initially the spores may be surrounded by a thin translucent membrane, but this soon breaks and smut spores are released into the crop. Within a few days all the spores are released and all that is left of the ear is the rachis.

The disease is normally assessed quite simply by estimating the percentage smutted ears in a crop.

(iv) *Disease cycle (Figure 6.23)*
The fungus overwinters as dormant mycelium in the embryo of contaminated seed. It is possible to detect only by excision and staining of the embryo. Almost 100% of contaminated seeds sown, whether in autumn or spring, will result in systematically infected plants. The dormant mycelium initially grows intracellularly to colonise seedlings. As the cereal plant matures, the fungal hyphae grow intercellularly within the plant, keeping up with the growing point. Leaves, roots and primordia of the ear are penetrated. Smut spores begin to develop in place of developing ovaries in the ear, and the ear emerges at around the time for anthesis in healthy plants. Spores are wind-blown to healthy cereal flowers where they germinate giving rise to basidia. The basidium produces haploid promycelia; these fuse to form dikaryotic hyphae which grow into the ovary. A dormant mycelium is established in the embryo of the developing grain. Warm humid conditions at anthesis appear to favour the disease.

Figure 6.21. Disease cycle of barley yellow mosaic virus (BaYMV). Vector: *Polymyxa graminis*.

(v) *Significance*

A combination of resistant varieties and fungicide seed treatments with activity against loose smut has dramatically reduced the prevalence of the disease in the UK. Seed certification schemes have also played an important part in control of the disease (see control). Crops will be rejected as seed crops if the maximum permitted levels of loose smut are exceeded. The yield of affected individual ears grown from contaminated seed will be zero. In crops grown from untreated home-saved seed, yield losses of up to 20%

Figure 6.22. Loose smut of wheat caused by *Ustilago nuda*.

have been recorded. However annual losses in the UK are on average probably less than 1%.

(vi) *Control*

Plant disease legislation, in the form of cereal seed certification schemes in the UK, has played an important part in the control of this potentially serious seed-borne disease. Loose smut contamination of seed crops must not exceed 0.5%. There are more rigorous standards of 0.2% contamination in crops grown for HVS (Higher Voluntary Standard). Hence the disease is kept to a minimum in all commercial cereal seed.

Disease resistance is available in both wheat and winter barley varieties. This resistance may be based on the simple biological phenomenon of the percentage of closed flowers in a variety. As smut spores can infect only via flowers, if flowers are never exposed to the atmosphere, the plant will escape infection.

Chemical control methods are available in the form of seed treatments which have systemic components included in their formulation. Systemicity is required in order to reach the mycelium in the embryo. Seed treatments are particularly useful for home-saved seed or varieties with poor resistance to loose smut (a high percentage of open flowers).

Products approved for the control of loose smut include the following:

Active ingredient	*Product example*
carboxin + imazalil + thiabendazole	Cerevax Extra (barley)
carboxin + phenylmercury acetate	Murganic RPB
carboxin + thiabendazole	Cerevax (wheat)
ethirimol + flutriafol + thiabendazole	Ferrax (barley)
fuberidazole + imazalil + triadimenol	Baytan IM (barley)
fuberidazole + triadimenol	Baytan

There is some evidence of resistance to carboxin in *U.nuda* (loose smut of barley).

Forecasting disease. Samples of seed can be checked by seed-testing stations for the presence of resting mycelium and the percentage of seeds contaminated will almost equal the percentage of plants affected should the seeds be sown.

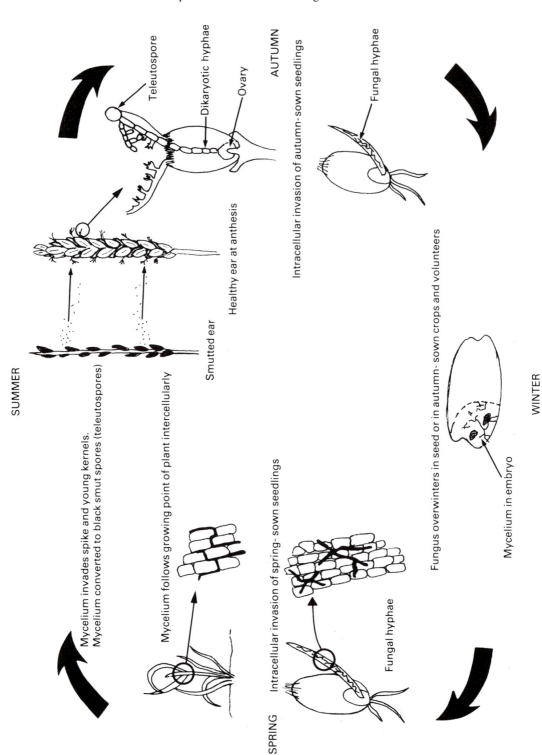

SUMMER

Teleutospore

Dikaryotic hyphae

Ovary

AUTUMN

Intracellular invasion of autumn-sown seedlings

Fungal hyphae

Healthy ear at anthesis

Smutted ear

Mycelium invades spike and young kernels.
Mycelium converted to black smut spores (teleutospores)

Mycelium follows growing point of plant intercellularly

WINTER

Fungus overwinters in seed or in autumn-sown crops and volunteers

Mycelium in embryo

SPRING   Intracellular invasion of spring-sown seedlings

Fungal hyphae

Ergot

(i) *Causal organism*
Fungus, *Claviceps purpurea*.

(ii) *Host range*
Wheat, barley, rye and many grass species. Infection of oats is rare.

(iii) *Symptoms*
The first symptoms of ergot are easily overlooked. Creamy to golden coloured droplets of honeydew exude from young florets in affected cereal ears. Saprophytic moulds such as *Cladosporium* spp. may colonise tissue coated with honeydew. The more obvious symptoms of ergot are the horn-shaped, purplish-black ergots or sclerotia which appear as individual replacements to grains. Ergots are usually around 10 mm in length, but may be up to 50 mm, especially in rye.

Assessment of ergot involves estimates of percentage of affected heads in a growing crop and/or checking for ergots in grain samples.

(iv) *Disease cycle (Figure 6.24)*
Ergots or sclerotia are hard, compact masses of fungal tissue which are the overwintering bodies of the fungus. Ergots overwinter either in the soil or as contaminants of grain. The ergot requires a dormant period at low temperatures before it germinates in the spring. The resulting perithecia release wind-blown ascospores. If ascospores alight on a susceptible floret, they quickly germinate and penetrate the ovary. Within a few days of infection, conidia are formed in honeydew which serves as secondary inoculum. Conidia are transferred to healthy florets by wind, splashing rain and insects. Honeydew formation and therefore build-up of disease in the crop is favoured by cool wet weather around flowering time. If pollination occurs before the inoculum reaches the flowers, infection is rare. As the crop matures, infected ovaries enlarge and are converted into sclerotia.

(v) *Significance*
Although ergots can reduce yield slightly by eliminating individual grains, ergot is more significant as a contaminant of grain samples. Ergots contain toxic alkaloids which can cause abortion. There are several historical reports of deaths in humans as a result of eating ergot-contaminated grain. Consequently, as with loose smut, there are maximum tolerances of ergot allowed in crops grown for seed.

Figure 6.23. Disease cycle of loose smut (*Ustilago* spp.).

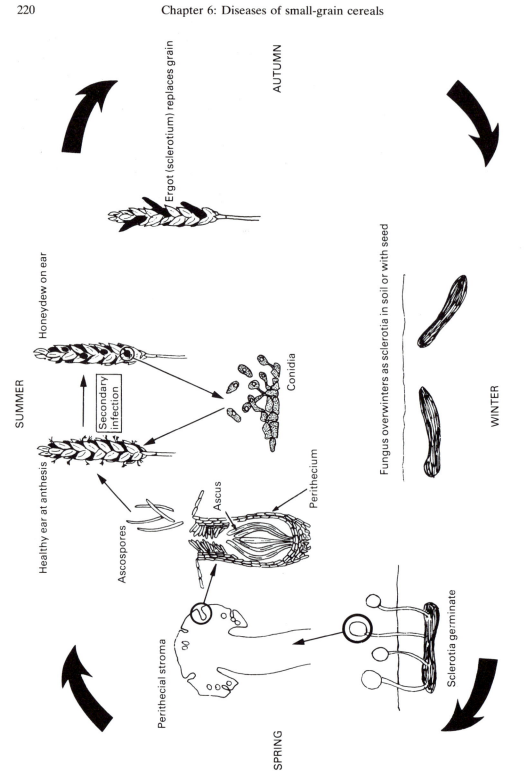

AUTUMN

Ergot (sclerotium) replaces grain

Honeydew on ear

SUMMER

Secondary infection

Conidia

Fungus overwinters as sclerotia in soil or with seed

WINTER

Healthy ear at anthesis

Ascus

Perithecium

Ascospores

Perithecial stroma

Sclerotia germinate

SPRING

Grain sold for human consumption should be ergot-free. Ergot has taken on an added significance in the recent production of F_1 hybrid cereals. Male-sterile lines in crossing programmes may have anthers exposed and unpollinated for a longer period than occurs naturally. The risks of ergot are, therefore, much higher than in normal seed production.

(vi) *Control*

Plant disease legislation which includes cereal seed certification schemes plays a major part in ensuring that seed is ergot-free in the UK. A maximum of three ergot pieces in 500 g seed is the tolerance for standard certified seed, and crops attempting to meet the HVS must be ergot-free.

Cultural control of ergot involves deep-ploughing of land in order to bury ergots to a depth in excess of 70 mm. This stops perithecial development. Rotations may also help to reduce the number of viable ergots in soil. Grass weeds should be eliminated to reduce the possibility of ergot spread from grass within the crop and on headlands. It is possible to remove ergots from grain by sieving or floating. Disease resistance and chemical control methods are not widely practised against ergot as other control methods are usually sufficient.

Sooty mould and black point

(i) *Causal organism*
Fungi such as species of *Alternaria* and *Cladosporium*.

(ii) *Host range*
Wheat, barley, oats and grasses.

(iii) *Symptoms*
Symptoms of sooty mould occur on ears of ripening cereals. Grey-black superficial fungal growth develops over the surface of the ear. Masses of spores may be released at harvesting. Deeper infection of grain by *Alternaria* spp. in particular can result in a darkening and shrivelling of the embryo end of the seed (black point).

(iv) *Disease cycle*
Figure 6.24. Disease cycle of ergot (*Claviceps purpurea*).

Fungi responsible for sooty mould are ubiquitous facultative parasites with numerous hosts. Infection occurs only if cereals have been stressed, particularly by other disease attacks such as take-all and eyespot. Alternatively, a delay in harvest as a result of wet weather may allow the fungi to infect the ripening plant.

(v) *Significance*

In itself, sooty mould is of little significance. It may result in extra dust during harvesting. Black point, on the comparatively rare occasions that it develops, does slightly worry millers of grain as, if samples have a high percentage of black point, flour may be slightly discoloured. Generally, the most significant aspect of sooty mould is the indication that there is another, perhaps more serious disease on the crop, which is weakening the plants.

(vi) *Control*

Control of sooty mould alone is not economically feasible. Fungicides aimed at controlling other more important diseases on the ear, e.g. *Septoria nodorum*, may reduce contamination.

Fusarium ear blight (scab)

(i) *Causal organism*

Fungi of genus *Fusarium*, including *Fusarium culmorum*, *F. avenaceum*, *F. graminearum*, *F. poae* and occasionally *F. nivale*.

(ii) *Host range*

Wheat and barley.

(iii) *Symptoms*

Individual spikelets become bleached as a result of infection. White mycelium, together with pink-orange spore masses, may also become visible. *Fusarium poae* produces different symptoms: pale lesions on glumes with a dark border.

(iv) *Disease cycle (Figure 6.8)*

The disease cycle of *Fusarium* ear blight is not well documented. It is generally assumed that ear infections arise primarily from spores splashed from infections at the stem base. There is little evidence to suggest systemic growth of the fungus in the plant. It is clear, however, that warm humid weather at flowering is necessary for disease development. Infection of the ears can result in seed contamination.

(v) *Significance*

The disease is very much weather-dependent. As such, severe epidemics of *Fusarium* ear blight probably occur only one year in ten in the UK. Yield losses after severe epidemics may be quite high because the grain is directly attacked. One per cent coverage of the ears may result in 1% yield loss.

(vi) *Control*

Disease resistance is available in winter wheat varieties (NIAB Farmers' Leaflet No. 8).

Chemical control may be financially worthwhile, particularly if there has been a period of wet weather around the time of flowering. Ideally, fungicides should be applied before the disease is seen. If more than 10% of the area of the ear is affected, control may be ineffective. Seed treatments may be effective against seed-borne inoculum (see *Fusarium* foot rot). Fungicide sprays approved for use against *Fusarium* ear blight include the following:

Active ingredient	*Product example*
carbendazim + maneb + sulphur	Bolda FL
maneb + zinc	Vassgro Manex
thiophanate-methyl	Cercobin Liquid

Strains of *F. culmorum* have been detected which are resistant to benzimidazole fungicides. Resistance is widespread in *F. nivale*.

Further reading

Asher, M. J. C. & Shipton, P. J. (ed.) (1981). *Biology and Control of Take-all*. London: Academic Press.

Association of Applied Biologists (1987). *Aspects of Applied Biology 15, Cereal Quality*. Wellesbourne, Warwick: Association of Applied Biologists.

Gair, R., Jenkins, J. E. E. & Lester, E. (1987). *Cereal Pests and Diseases*. Ipswich: Farming Press.

Gareth Jones, D. & Clifford, B.C. (1983). *Cereal Diseases: Their Pathology and Control*. Chichester: Wiley-Interscience.

King, J. E., Cook, R. J. & Melville, S. C. (1983). A review of *Septoria* diseases of wheat and barley. *Annals of Applied Biology*, **103**, 345–73.

Mathre, D. E. (ed.) (1982). *Compendium of Barley Diseases*. St Paul: American Phytopathological Society.

Ministry of Agriculture, Fisheries and Food (1986). *The Use of Fungicides and Insecticides on Cereals 1986*. Booklet 2257 (86). Alnwick, Northumberland: MAFF Publications.

National Institute of Agricultural Botany (1989) *Recommended Varieties of Cereals* 1989. Farmers' Leaflet No. 8. Cambridge: NIAB publication.

Wiese, M.V. (1987). *Compendium of Wheat Diseases*. St Paul: American Phytopathological Society.

7 Diseases of oilseed rape

Oilseed rape has rapidly increased in popularity in the EC since the early 1970s directly as a result of large subsidies on vegetable oil. The popularity of oilseed rape is now beginning to decline in the UK following the reduction of subsidies and the imposition of higher quality standards on oil and meal produced from the crop. Rapeseed oil must now contain only very small amounts of erucic acid, and meal very small amounts of glucosinolates, both potentially harmful chemicals. Special 'double-low' varieties are available to farmers to meet these demands, but they are not as productive as older 'single-low' (low in erucic acid) varieties.

There has been a succession of serious disease epidemics in oilseed rape during its rise in popularity in the UK. Stem canker posed a major threat to production during the late 1970s and there was an epidemic of *Alternaria* pod spot during 1981. The rapid breeding of resistant varieties and the development of fungicides have, respectively, rescued the crop from these problems. Light leaf spot is probably the most widespread damaging disease of oilseed rape in the UK at present, although *Sclerotinia* stem rot is of growing importance in southern UK. All pathogens on oilseed rape can also infect vegetable brassicas.

The use of fungicides on oilseed rape has increased in recent years. However, the practical difficulties involved in taking machinery through such a dense and comparatively tall crop make it desirable to apply only the minimum number of sprays. Hence a single application of a broad-spectrum fungicide, applied either at stem extension (aimed primarily at the control of light leaf spot) or between mid and late flowering (for control of *Alternaria*), is a common practice.

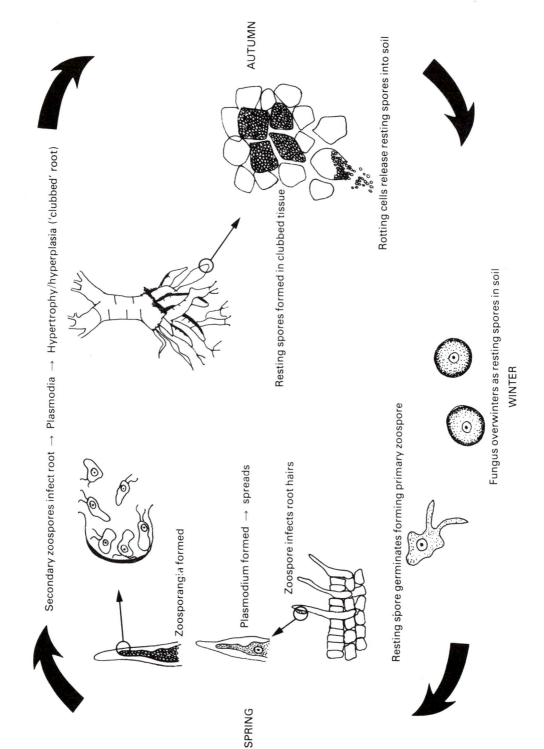

SUMMER

AUTUMN

Secondary zoospores infect root → Plasmodia → Hypertrophy/hyperplasia ('clubbed' root)

Resting spores formed in clubbed tissue

Rotting cells release resting spores into soil

Zoosporangia formed

Plasmodium formed → spreads

Zoospore infects root hairs

Resting spore germinates forming primary zoospore

Fungus overwinters as resting spores in soil

WINTER

SPRING

Clubroot

(i) *Causal organism*
Fungus, *Plasmodiophora brassicae.*

(ii) *Host range*
All members of the Cruciferae are susceptible, including cabbage, Brussels sprout, turnip, kale, cauliflower, broccoli and radish, as well as cruciferous weeds such as charlock and shepherd's purse.

(iii) *Symptoms (Figure 4.2, p.71)*
Clubroot affects the roots of oilseed rape. Clubs, or irregular swellings, are formed. The disease is sometimes called 'finger and toe' disease, as lateral rootlets as well as the main root become swollen. The capacity of infected plants to take up water and nutrients may be seriously impaired and hence further typical symptoms of the disease include stunting and wilting. If root swellings are cut open, internal tissue may have a marbled appearance.

(iv) *Disease cycle (Figure 7.1)*
The fungus survives in the soil as resistant and highly durable resting spores which are released from infected decaying root tissue. They can probably persist in the soil for 20 years or more. Under suitable conditions, which probably include increases in temperature and moisture in the soil, the resting spores germinate to produce single biflagellate primary zoospores which swim in soil moisture. On contact with a susceptible host root, the zoospores encyst on the surface of root hairs or epidermal cells and penetrate the cell wall. Plasmodia are then formed within cells. After growth of the plasmodia, zoosporangia are formed within cells and these give rise to secondary zoospores. The precise function of secondary zoospores is uncertain; it is thought that they may penetrate deeper into root tissue where secondary plasmodia are formed. This appears to coincide with the major phase of club formation. Towards the end of the growing season, plasmodia undergo meiosis, forming more resting spores which are released into soil when tissue decays. The disease causes most problems on poorly drained acid soils.

(v) *Significance*
At present, the disease is common only in areas with a history of brassica growing, e.g. Lancashire (UK). Clubroot rarely kills oilseed rape plants, but widespread and severe attacks can reduce

Figure 7.1. Disease cycle of clubroot (*Plasmodiophora brassicae*).

yield significantly. Precise economic losses are difficult to determine as mild attacks often remain undetected in oilseed rape.

(vi) *Control*

Cultural control methods offer the best solution to the problem of clubroot. Rotation may be of some help although periods between susceptible crops may have to be as long as 8 years. Cruciferous weeds should be eliminated and drainage improved if necessary. An increase in soil pH, to pH 6.5 or above by liming, will also reduce the disease.

Clubroot is the only disease of any significance which affects oilseed rape roots.

Sclerotinia stem rot

(i) *Causal organism*
Fungus, *Sclerotinia sclerotiorum*.

(ii) *Host range*
The fungus has one of the widest host ranges of any pathogen. Most agricultural crops including other Cruciferae, Solanaceae (tomato, potato), Compositae (lettuce, sunflower), Leguminosae (pea, bean), Umbelliferae (carrot), Chenopodiaceae (beet) and a number of weed species may be infected.

(iii) *Symptoms (Figure 7.2)*
Stem rot symptoms are usually seen in the crop from mid-May onwards. Bleached areas, often associated with leaf axils or branch points, develop from 20–50 cm above soil level. The lesions may enlarge rapidly giving rise to an almost entirely bleached stem. If the stem is split open, hard black resting bodies, sclerotia, are frequently found within the stem cavity. Affected plants ripen prematurely and stems may crack, resulting in lodging.

(iv) *Disease cycle (Figure 7.3)*
The fungus survives as sclerotia either in the soil or as loose unobtrusive contaminants of seed (Figure 7.2). The latter may be important in the introduction of the disease into new planting areas. In the spring sclerotia, which are buried in the top 3–5 cm of soil, germinate to produce apothecia. Ascospores released from the apothecia are wind-blown to oilseed rape stems and petals. Pre-requisites for ascospore infection of stems are the primary

colonisation of oilseed rape petals and warm wet weather. When petals fall, many adhere to the stems where they may rot, thus providing a food source for the ascospores and facilitating their entry into the stem tissue. There is also evidence of direct mycelial infection at soil level from sclerotia. Optimum weather conditions for ascospore germination are temperatures around 20 °C and relative humidities above 95%. Sclerotia develop in stem cavities towards the end of the season. At harvest, these are either taken up

Figure 7.2. Stem rot of oilseed rape caused by *Sclerotinia sclerotiorum*. Note black sclerotia emerging through stem tissue. (S. Mitchell, Imperial College.)

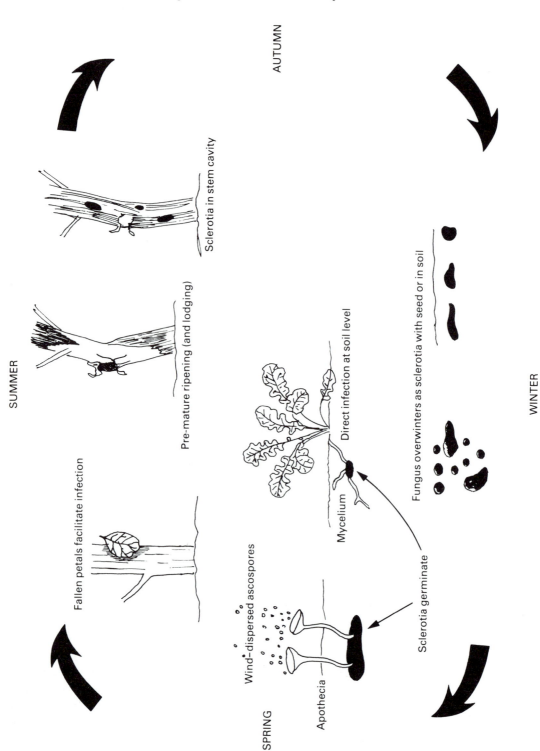

AUTUMN

SUMMER

Sclerotia in stem cavity

Pre-mature ripening (and lodging)

Fallen petals facilitate infection

Direct infection at soil level

Fungus overwinters as sclerotia with seed or in soil

WINTER

Mycelium

Wind-dispersed ascospores

Sclerotia germinate

Apothecia

SPRING

with the seed or fall to the soil, where they may remain viable for eight years or more.

(v) *Significance*

Epidemics of *Sclerotinia* are sporadic at present but the disease is a serious potential threat to oilseed rape production in southern UK. Individual diseased plants will yield very little and losses in excess of 20% have been reported in severe disease attacks.

(iv) *Control*

Cultural control includes deep ploughing, an attempt to bury sclerotia in soil or stubble so deep that apothecia cannot reach the soil surface. Rotations which avoid long runs of susceptible crops should be adopted. Machinery should be cleaned after harvesting infected crops and seed inspected; seed contaminated by sclerotia should be avoided.

Disease resistance in currently grown oilseed rape varieties is low. Both early and late-flowering varieties may escape disease because there are no petals available for colonisation during warm humid spring weather.

Chemical control methods are available, but it is unlikely that routine prophylactic spraying is economically worthwhile. However, once disease symptoms are evident, fungicides will probably have little effect on infected plants; they will protect only uninfected crops. Fungicides applied at early to mid-flowering in fields with a history of the disease are likely to be most cost-effective.

Fungicides approved for the control of *Sclerotinia* include:

Active ingredient	Product example
carbendazim + prochloraz	Sportak Alpha
iprodione	Rovral Flo
prochloraz	Sportak
thiophanate-methyl	Cercobin Liquid
vinclozolin	Ronilan FL

Figure 7.3. Disease cycle of stem rot (*Sclerotinia sclerotiorum*).

Forecasting disease would be a great advantage in the rational use of fungicides for control of *Sclerotinia*. The basis for forecasting should involve assessment of amount of inoculum in fields, and timing of ascospore release and petal fall; together with temperature and humidity requirements for infection and disease development.

Phoma leaf spot and stem canker

(i) *Causal organism*
Fungus, *Phoma lingam* (*Leptosphaeria maculans*).

(ii) *Host range*
Cruciferae.

(iii) *Symptoms (Figure 7.4)*
The first symptoms of *Phoma* leaf spot are bleached, usually circular spots containing dark pycnidia which may be seen on cotyledons and young leaves. In early summer the fungus causes the more serious problem of cankers on the stem. These frequently develop at the base of the stem close to points of attachment of lower leaves (leaf scars). Brown to black patches are the first symptoms of stem canker to develop. These are followed by brown cankers which may deeply penetrate the stem, resulting in lodging and premature ripening of the pods. Black pycnidia may again be visible. The fungus can also infect the pods, giving rise to symptoms similar to those produced on leaves.

Figure 7.4. *Phoma* leaf spot of oilseed rape caused by *Phoma lingam* (*Leptosphaeria maculans*) Note dark pycnidia in lesions. (Schering Agriculture.)

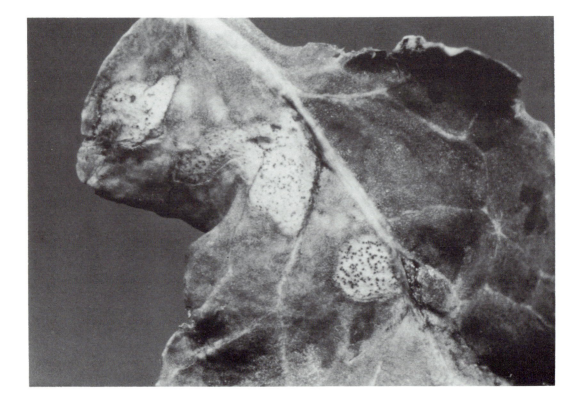

(iv) *Disease cycle (Figure 7.5)*

The two main sources of overwintering inoculum are contaminated seed and oilseed rape stubble. The former may introduce the disease to new cropping areas, whereas the latter results in infection of the existing crop. The fungus overwinters on stubble as pseudothecia. These release ascospores in the autumn or spring which result in the primary leaf spots containing pycnidia. The pycnidia release spores which may be rain-splashed in the crop giving rise to further leaf spots early in the season. Spores may also be washed onto the stem where penetration of old leaf scars occurs and cankers develop. Additionally, the fungus is able to colonise the stem by systemic growth within the xylem, starting from the leaf lesions. Spores may also be rain-splashed to the pods where seed infection is possible. After harvest, the fungus rapidly colonises stem and root tissue and pseudothecia may be formed on debris. These may be extremely durable and have been reported as viable after 10 years on stubble in the soil. Precise weather conditions which favour disease development have not been defined, although warm wet conditions are understood to be favourable.

(v) *Significance*

Major epidemics of stem canker occurred in a number of areas of the world during the 1970s. The disease threatened the production of oilseed rape in some areas and yield losses of up to 60% were reported. However, the disease has declined in importance recently because of the introduction of disease-resistant varieties.

(vi) *Control*

Cultural control is very important in dealing with *Phoma*. Contaminated stubble should be disposed of quickly after harvest by burning, ploughing or deep cultivation. Crop rotations and isolation from other oilseed rape crops and brassicas will also help.

Disease resistance in oilseed rape varieties is almost essential if a variety is to succeed commercially. Disease resistance ratings are given in the NIAB Farmers' Leaflet No. 9, *Recommended Varieties of Oilseed Rape* for stem canker, light leaf spot and downy mildew.

Chemical control of leaf spot and stem canker may be undertaken by seed treatment and foliar sprays in autumn and/or early spring.

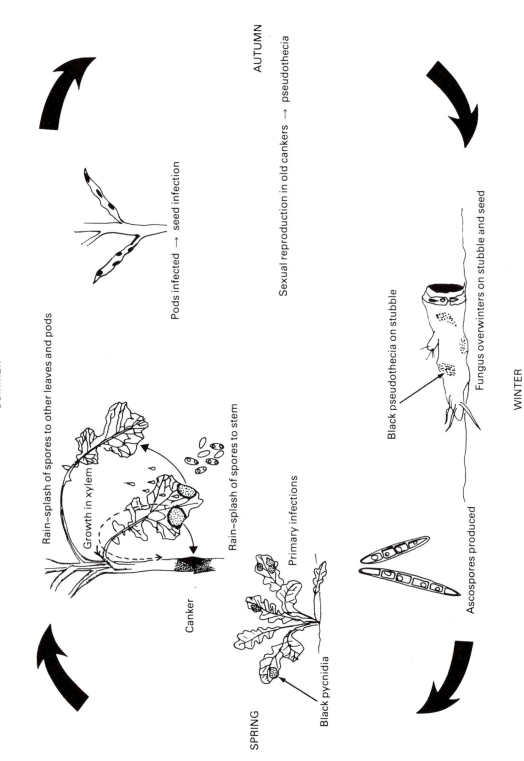

AUTUMN

Sexual reproduction in old cankers → pseudothecia

Pods infected → seed infection

SUMMER

Rain-splash of spores to other leaves and pods

Growth in xylem

Rain-splash of spores to stem

Canker

Black pseudothecia on stubble

Fungus overwinters on stubble and seed

WINTER

Primary infections

Ascospores produced

Black pycnidia

SPRING

The latter have given rather inconsistent yield results. Fungicides approved for control of the disease include:

Active ingredient	*Product example*
carbendazim + gamma-HCH + thiram	Gammalex Liquid (seed treatment)
carbendazim + prochloraz	Sportak Alpha
carboxin + gamma-HCH + thiram	Vitavax RS Flowable (seed treatment)
fenpropimorph + gamma-HCH + thiram	Lindex-Plus FS (seed treatment)
gamma-HCH + thiabendazole + thiram	New Hysede FL (seed treatment)
iprodione	Rovral WP (seed treatment)
prochloraz	Sportak
thiabendazole	Tecto 60% WP (seed treatment)

Botrytis grey mould

(i) *Causal organism*

Fungus, *Botrytis cinerea* (*Botryotinia fuckeliana*).

(ii) *Host range*

Numerous hosts.

(iii) *Symptoms*

All above-ground parts of oilseed rape can be attacked by *B. cinerea*. However, some form of tissue damage is a pre-requisite for infection to occur. The disease is most common on leaves, where grey-brown irregular *Botrytis* lesions can be seen early in the season associated with senescence and damage caused by fertilisers, frost and hail. On mature leaves, similar *Botrytis* lesions are frequently found associated with petals adhering to the leaf surface. *Botrytis* can also cause bleached lesions on stems, again usually associated with some form of damage such as cracks or broken branches. During humid weather, typical grey mould may be seen on leaf and stem surfaces. Pest damage on pods frequently results in *Botrytis* infection; pods may decay, become dehydrated and again become covered with a grey mouldy growth.

Figure 7.5. Disease cycle of leaf spot and stem canker (*Phoma lingam* = *Leptosphaeria maculans*).

(iv) *Disease cycle*

Botrytis cinerea is a ubiquitous pathogen which exists all year round on senescent, decaying or dead plant tissue. The typical grey mould consists of hyphae, with conidiophores bearing masses of light, wind-blown spores. If spores alight on damaged oilseed rape tissue, they can penetrate the surface and, depending on the severity of damage, colonise large areas of tissue. Temperatures of 10–15 °C together with high relative humidities favour disease development.

(v) *Significance*

The disease is common on leaves but severe epidemics are rare. Stem lesions cause premature ripening of distal portions of raceme which is alarming in appearance but rarely significant.

(vi) *Control*

Specific control measures are seldom justified. Damage to crops should be avoided where possible.

Chemical control measures are available. It is unlikely that fungicide application specifically for *Botrytis* would be economically justifiable, but broad-spectrum oilseed rape fungicides applied to control other more economically important diseases such as *Alternaria* are quite effective against *Botrytis*. Fungicides approved for control of *Botrytis* include:

Active ingredient	Product example
benomyl	Benlate
carbendazim + chlorothalonil	Bravocarb
carbendazim + maneb + sulphur	Bolda FL
carbendazim + prochloraz	Sportak Alpha
chlorothalonil	Bravo 500
iprodione	Rovral Flo
iprodione + thiophanate-methyl	Compass
prochloraz	Sportak
thiophanate-methyl	Cercobin Liquid
vinclozolin	Ronilan FL

Light leaf spot

(i) *Causal organism*

Fungus, *Pyrenopeziza brassicae* (*Cylindrosporium concentricum*)

(ii) *Host range*

The disease is restricted to members of the genus *Brassica* (family Cruciferae).

(iii) *Symptoms (Figure 7.6)*

The disease may be found on all aerial parts of the plant. On the leaves, light green bleached or speckled areas develop, particularly in early spring. These symptoms can very easily be confused with mechanical or frost damage or fertiliser scorch. However, during wet weather in particular, small white spore masses (acervuli) develop scattered in and around the light leaf spot lesions. If lesions

Figure 7.6. Light leaf spot of oilseed rape caused by *Pyrenopeziza brassicae* (*Cylindrosporium concentricum*) (i) On leaves; note white spore masses (acervuli). (ii) On stems; note black speckling. (C. J. Rawlinson, Rothamsted.)

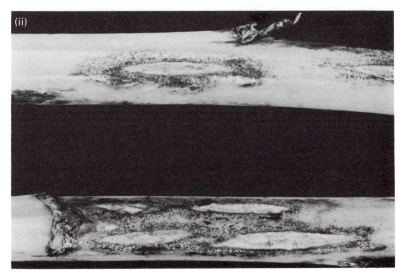

are caused by the fungus rather than damage, acervuli will develop overnight on leaves put in a polythene bag with damp tissue. On stems, infections result in fawn elongated lesions with black speckled edges and eventually a split epidermis. The disease may spread to buds, which can be killed, and later to pods, which look mealy as a result of acervuli development and may become twisted.

(iv) *Disease cycle*

The fungus overwinters mainly as resting mycelium and conidia on crop debris, volunteers, early autumn-sown crops and other brassicas. Conidia may also overwinter on debris carried with the seed, and on the seed itself. Free water is necessary for infection to occur, together with temperatures between 5 and 15 °C; water also disperses conidia. Most infections take place during mild, wet autumn and early spring weather. Dry weather inhibits further spread of the disease.

(v) *Significance*

Light leaf spot is a very common disease on oilseed rape in the spring and arguably the most important disease of the crop in the UK at present. The disease can reduce yield significantly during severe attacks by tissue death of leaves, buds, flowers and pods. Yield losses of 50% have been reported in individual crops.

(vi) *Control*

Cultural control methods include disposal of contaminated stubble, together with eradication of volunteers, isolation of oilseed rape from other brassicas and the use of crop rotation.

Disease resistance is very important in oilseed rape varieties. One of the first, most popular varieties to be grown in the UK, Jet Neuf, was very susceptible to light leaf spot and caused the farmer many problems. Most of the current varieties have more resistance to light leaf spot.

Chemical control methods are available. Recommendations of spray timing and disease thresholds vary depending upon the fungicide. Generally, an early spray (at onset of stem extension) will prove useful. If the disease is observed later in the season, a further spray at the end of flowering may also be applied. Fungicides approved for the control of light leaf spot include the following:

Active ingredient	Product example
benomyl	Benlate
carbendazim	Bavistin, Derosal Liquid, Stempor DG
carbendazim + chlorothalonil	Bravocarb
carbendazim + mancozeb	Kombat
carbendazim + maneb	Squadron
carbendazim + maneb + sulphur	Bolda FL
carbendazim + prochloraz	Sportak Alpha
iprodione + thiophanate-methyl	Compass
prochloraz	Sportak
propiconazole	Tilt 250 EC, Radar
thiophanate-methyl	Cercobin Liquid
vinclozolin	Ronilan FL

There have been reports of resistance to benzimidazole fungicides in. *P. brassicae*. Although the resistance is not widespread as yet, it may become so in the near future.

Dark leaf and pod spot (*Alternaria*)

(i) *Causal organism*
Fungi, mainly *Alternaria brassicae* and, to a lesser extent *Alternaria brassicicola*.

(ii) *Host range*
Many cruciferous plants.

(iii) *Symptoms (Figures 4.1 (p. 70), 7.7)*
Symptoms of *Alternaria* may be seen on oilseed rape leaves throughout the growing season. The first signs of infection are small dark brown flecks on leaf tissue, often surrounded by a chlorotic halo. These spots enlarge to produce characteristic 'target spot' lesions consisting of dark brown circular areas containing brown concentric rings. These lesions may also be surrounded by a chlorotic zone. *Alternaria* symptoms on pods of oilseed rape consist of dark irregular lesions which can result in premature ripening, splitting and seed shed prior to harvest.

(iv) *Disease cycle (Figure 7.8)*
The fungi overwinter on contaminated seed, crop stubble and debris, volunteers, autumn-sown crops and overwintering vegetable brass-

AUTUMN

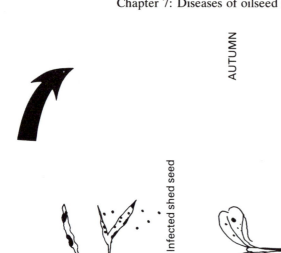

Infected shed seed

Infected pods

Dark spots on newly emerged leaves

u

SUMMER

WINTER

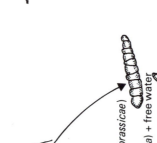

Target spots

Epidemic
Temp. 17–24°C (A. brassicae)
23–35°C (A. brassicicola) + free water

Spores

Fungi overwinter on seed, stubble, debris, volunteers
autumn-sown crops, vegetable brassicas

SPRING

icas. As temperatures rise in the spring, large spores are produced on infected tissue, which are dispersed short distances by rain-splash and longer distances on air-currents. Free water is necessary for germination and penetration of both *Alternaria* species but the two have slightly different temperature optima; the most common species, *A. brassicae*, from 17 to 24 °C, *A. brassicicola*, 25–35 °C. Vertical dispersion of spores results in symptoms on pods and frequently colonisation of seeds. Warm wet weather during early summer is ideal for secondary infections to occur. Contaminated seed, shed prior to harvest, can give rise to infected volunteers.

Figure 7.7. Dark pod spot of oilseed rape caused by *Alternaria brassicae*. (Schering Agriculture.)

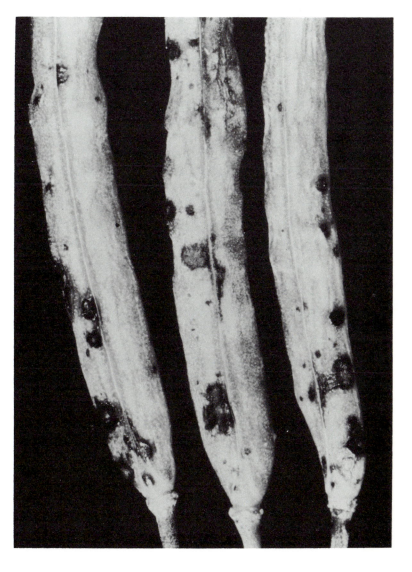

Figure 7.8. Disease cycle of dark leaf and pod spot (*Alternaria brassicae*, *A. brassicicola*).

(v) *Significance*

In Europe, *Alternaria* is widespread and arguably the most important disease of oilseed rape. Epidemics may occur as frequently as every two years and yield losses of up to 60% have been reported, mainly as a result of disease on the pods.

(vi) *Control*

Cultural control methods are as for *Phoma* leaf spot and stem canker.

Chemical control methods involve both fungicide seed treatments and foliar sprays. It is most critical to protect the pods from infection. In some seasons, it may even be economically worthwhile to spray twice, at mid-flowering and 3–4 weeks later.

Fungicides approved for the control of *Alternaria* include the following:

Active ingredient	*Product example*
carbendazim + chlorothalonil	Bravocarb
carbendazim + maneb + sulphur	Bolda FL
carbendazim + prochloraz	Sportak Alpha
fenpropimorph + gamma-HCH + thiram	Lindex-Plus FS (seed treatment)
ferbam + maneb + zineb	Trimanzone
iprodione	Rovral Flo
iprodione + thiophanate-methyl	Compass
maneb	Ashlade Maneb Flowable
maneb + zinc	Manex
prochloraz	Sportak
propiconazole	Tilt 250 EC, Radar
vinclozolin	Ronilan FL

Forecasting disease would be particularly advantageous in the effective use of fungicides for control of *Alternaria*. Wet weather conditions before pod set could be a basis for forecasting the disease.

Downy mildew

(i) *Causal organism*
Fungus, *Peronospora parasitica*.

(ii) *Host range*
Cruciferae.

(iii) *Symptoms*
Symptoms of downy mildew are most commonly seen on leaves of oilseed rape in the autumn and early spring. Angular chlorotic patches occur on the upper side of the leaf and, associated with these patches, a white or grey downy fungal growth is evident on the underside of the leaf. During dry weather, lesions may become necrotic and dry out. However, during periods of cool wet weather, the disease progresses and it is possible for stems and pods to become infected later in the season.

(iv) *Disease cycle (Figure 7.9)*
The pathogen overwinters as resting spores (oospores) either on plant debris or in the soil. During wet periods from the autumn through to the spring, if temperatures are mild, oospores germinate producing sporangia, which are thought to initiate infections. Sporangia germinate on plant tissue and penetrate directly or via stomata. After 3–4 days at around 12 °C with high relative humidity, sporangiophores bearing more sporangia are formed. These are dispersed passively by wind and rain, and possibly actively by hygroscopic twisting of the sporangiophore in response to changes in relative humidity. Towards the end of the cycle of disease, oospores are formed in dying tissue, which will later reach the soil.

(v) *Significance*
Downy mildew is the most common disease on young oilseed rape plants in the autumn. Severe attacks coupled with winter frosts can reduce plant populations. However, plants usually grow away from the disease and epidemics of downy mildew are rare after early spring.

(vi) *Control*

Disease resistance to downy mildew is available in currently grown oilseed rape varieties (NIAB Farmers' Leaflet No. 9, *Recommended Varieties of Oilseed Rape*.) It is possible that such resistance will not prove durable.

Chemical control methods are also available, but there is some debate as to their economic viability. If applied, early spraying,

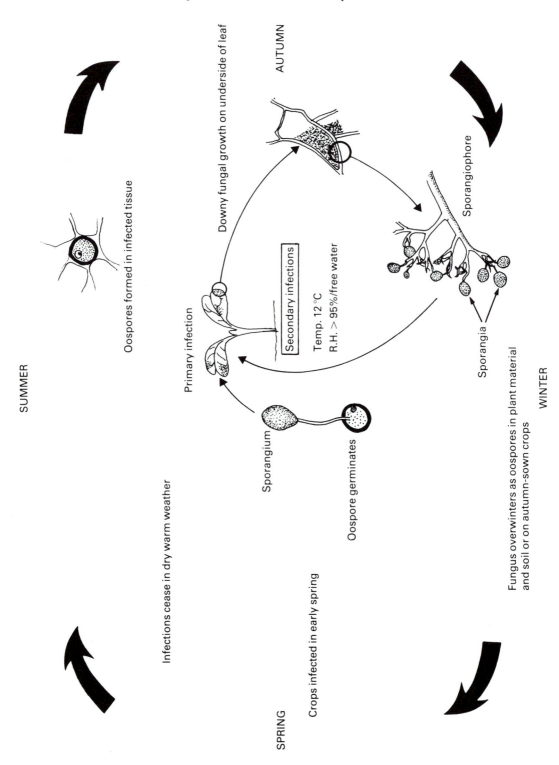

before the end of November, is recommended. Fungicides approved for control of downy mildew include the following:

Active ingredient	Product example
carbendazim + mancozeb	Kombat
chlorothalonil	Bravo 500
chlorothalonil + metalaxyl	Folio 575 FW
mancozeb	Dithane 945
mancozeb + metalaxyl	Fubol 58 WP
maneb	Ashlade Maneb Flowable
maneb + zinc	Manex
manganese zinc ethylenebisdithio- carbamate + ofurace	Patafol Plus
zineb-ethylenethiuram disulphide complex	Polyram

Powdery mildew

(i) *Causal organism*
Fungus, *Erysiphe cruciferarum*.

(ii) *Host range*
Cruciferae.

(iii) *Symptoms*
Initial symptoms are fairly small star-shaped patches of white mycelium on leaves. Symptoms may progress to form an almost total white superficial covering of the leaf surface. Masses of wind-blown spores are released if infected leaves are disturbed. The disease occurs on both sides of oilseed rape leaves and it can spread to the stems and pods. Unlike downy mildew, there is little or no chlorosis associated with fungal colonisation initially, although this may develop later.

(iv) *Disease cycle*
The fungus survives the winter mainly as resting mycelium in volunteers, early autumn-sown crops and other crucifers. Cleistothecia are rarely formed on oilseed rape and are probably of little significance. As temperatures rise during the spring, conidia are produced which are wind-dispersed throughout the crop, initiating fresh mildew colonies. A mildew epidemic is most likely in early summer when temperatures are between 17 and 20 °C and relative humidity is high. Mildew is inhibited above 20 °C.

Figure 7.9. Disease cycle of downy mildew (*Peronospora parasitica*).

(v) *Significance*

The disease is very seasonal. Severe epidemics, which can reduce yield significantly, are rare.

(vi) *Control*

Cultural control methods include eradication of volunteers and weed crucifers. Isolation of oilseed rape from other brassicas will be of benefit as will crop rotations. Autumn-sown oilseed rape crops should not be planted too early. In the UK, ADAS advise that sowing before 20 August can encourage both powdery mildew and *Alternaria* infection and should therefore be avoided.

Disease resistance to powdery mildew does exist in oilseed rape varieties, but there are insufficient differences between varieties at present for resistance ratings to be published by the NIAB.

Chemical control of powdery mildew may be justified in fields with a history of the disease. Application during March of sulphur as Thiovit, Solfa or Magnetic 6, for example, is likely to reduce powdery mildew in the spring.

Ring spot

(i) *Causal organism*
Fungus, *Mycosphaerella brassicicola*.

(ii) *Host range*
Most *Brassica* species.

(iii) *Symptoms*
Ring spot can develop on all aerial parts of the plant, but is most commonly seen on leaves of oilseed rape. Small dark spots, similar to the initial stages of *Alternaria* infection, are the first observable symptoms. Lesions then develop as circular brown necrotic areas containing concentric rings of dark pseudothecia or pycnidia. Separate lesions may be bounded by a chlorotic margin. During severe disease attacks, leaves may become yellow and crack.

(iv) *Disease cycle*
The fungus overwinters primarily as pseudothecia and pycnidia on oilseed rape debris, stubble, volunteers and vegetable brassicas. Seed-borne infection has also been reported. During spring,

wind-dispersed ascospores released from pseudothecia are the primary initial source of inoculum. The role of pycnidia in disease spread is unclear. Temperatures of around 15 °C, together with high relative humidity, tend to favour disease development.

(v) *Significance*
As yet, ringspot is of minor importance on oilseed rape and much more of a problem on vegetable brassicas in the UK. The few recent severe outbreaks suggest that it may assume a greater importance in the near future.

(vi) *Control*
Cultural control methods, as for *Phoma* leaf spot and stem canker, would be useful in reducing the disease.

Chemical control methods are probably unjustified with present levels of the disease and there are no approved fungicides for control of ringspot.

White leaf spot

(i) *Causal organism*
Fungus, *Pseudocercosporella capsellae*.

(ii) *Host range*
Brassica spp. and some cruciferous weeds.

(iii) *Symptoms*
Fairly small bleached spots occur mainly on the leaves although they can also be found on pods. As lesions age, they may develop grey centres. The symptoms are difficult to distinguish from fertiliser or herbicide scorch. However, chemical damage is usually uniformly spread over leaf surfaces whereas white leaf spot lesions often occur in specific areas.

(iv) *Disease cycle*
The fungus probably overwinters as resting mycelium or sclerotia on oilseed rape stubble, volunteers, weeds and other *Brassica* species. Cool wet weather favours disease development. It is likely that spores are splash-dispersed in the crop.

(v) *Significance*

As yet, the disease is only of minor significance in the UK although it is of increasing importance in the south-west of England and Wales. In Australia, however, very susceptible lines of oilseed rape have suffered almost complete defoliation as a result of white leaf spot attacks.

(vi) *Control*

Specific control measures are unlikely to be necessary for control of the disease in the UK. Good crop hygiene will normally help to contain the disease. Prochloraz (Sportak) and carbendazim + prochloraz (Sportak Alpha) applied in the spring are approved for control of the disease.

Virus diseases

Viruses which have been reported in oilseed rape in the UK include beet western yellows (BWYV), cauliflower mosaic (CaMV), turnip mosaic (TuMV), broccoli necrotic yellows (BNYV) and beet mild yellowing (BMYV). The main vectors of these viruses are the aphid species *Myzus persicae* and *Brevicoryne brassicae*. The effects, incidence and significance of virus infection in oilseed rape have yet to be determined, but Walsh & Tomlinson (1985) suggested in a limited survey that cauliflower mosaic and turnip mosaic were the two most common viruses, although their incidence was low. They also found that yields were reduced significantly as a result of infection by these two viruses. According to Smith & Hinckes (1985), 100% BWYV infection in experimental plots of oilseed rape reduced yield of oil by over 13% compared with plots with 18% virus infection. There is clearly potential for problems resulting from virus infection in oilseed rape. The diversity of potential hosts for many of these viruses, which for BWYV includes sugar beet, lettuce, sunflower, turnip and many *Brassica* spp. also gives cause for concern.

A schematic representation of disease symptoms on oilseed rape leaves is given in Figure 7.10.

Further reading

Davies, J. M. L. (1986). Diseases of oilseed rape. In *Oilseed Rape*, ed. D. H. Scarisbrick & R. W. Daniels, pp. 195–236. London: Collins.

Dixon, G. R. (1981). *Vegetable Crop Diseases*. London: Macmillan.

Ministry of Agriculture, Fisheries and Food (1984). *Control of Pests and Diseases of Oilseed Rape*. ADAS Booklet 2387 (84) Alnwick, Northumberland: MAFF publications.

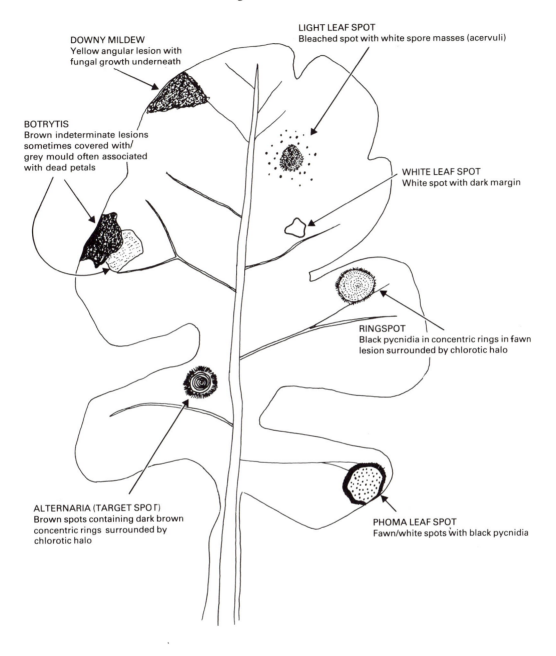

DOWNY MILDEW
Yellow angular lesion with
fungal growth underneath

LIGHT LEAF SPOT
Bleached spot with white spore masses (acervuli)

BOTRYTIS
Brown indeterminate lesions
sometimes covered with
grey mould often associated
with dead petals

WHITE LEAF SPOT
White spot with dark margin

RINGSPOT
Black pycnidia in concentric rings in fawn
lesion surrounded by chlorotic halo

ALTERNARIA (TARGET SPOT)
Brown spots containing dark brown
concentric rings surrounded by
chlorotic halo

PHOMA LEAF SPOT
Fawn/white spots with black pycnidia

Figure 7.10. Schematic representation of disease symptoms on oilseed rape leaves.

National Institute of Agricultural Botany (1989). Farmers' Leaflet No. 9, *Recommended Varieties of Oilseed Rape*. Cambridge: NIAB publication.

Priestley, R. H. & Knight, C. K. (1985). *Diseases of Oilseed Rape and Fodder Brassicas*. Cambridge: NIAB publication.

Ward, J. T., Basford, W. D., Hawkins, J. H. & Holliday, J. M. (1985). *Oilseed Rape*. Ipswich: Farming Press.

8 Diseases of field peas and beans

The area of field peas and beans grown in the UK is increasing. From a total area of 64 000 ha in 1982, field peas and beans occupied 195 000 ha in 1987 (England and Wales). The EC has a strong commitment to pulses and has consistently increased its support in the form of subsidies over the last few years. There is a strong demand for the high quality protein feed produced from pea and bean crops.

Several important pathogens of peas and beans are seed-borne and seed treatments are becoming increasingly popular.

Foot and root rot

(i) *Causal organism*
Fungi, including a number of *Fusarium* species, especially *F. solani* f.sp. *pisi* (peas) and *F. solani* f.sp. *phaseoli* (beans). *Aphanomyces euteiches* f.sp. *pisi* (peas), *A. euteiches* f.sp. *phaseoli* (beans) and *Phoma medicaginis* var. *pinodella* (peas) have also been implicated.

(ii) *Host range*
The *formae speciales* and var. pathogens listed above are comparatively specific to peas and beans. Other fungi such as non-specialised *Fusarium* spp. causing the disease attack a very wide range of hosts.

(iii) *Symptoms*
Most of the above pathogens can cause damping-off in peas and beans, especially if the weather is cold and wet and germination is slow. More commonly, however, symptoms are seen in mature plants. Wilting may occur after warm periods and patches of stunted plants can be seen in the crop. Close examination of infected plants

reveals a dry rotting of the upper tap root and brown rotten roots. Infections caused primarily by *Fusarium* spp. can lead to pink-orange spore masses developing on the surface of the stem base, or in the lumen.

(iv) *Disease cycle*

This varies according to the pathogen involved. Most of the pathogens are soil-borne and can form specialised resting structures (e.g. chlamydospores in *Fusarium* spp.). Sporulation of fungi on aerial plant parts can lead to spread of the disease in the crop. Generally warm wet weather favours disease development and there are more problems in heavy, poorly drained soils. Soil compaction or other plant stresses also favour the disease.

(v) *Significance*

The disease is fairly widespread in the UK although it is only a particular problem in certain fields during very wet seasons.

(vi) *Control*

Cultural control involving good crop rotations helps to reduce the disease. Improvement in soil drainage and structure also helps.

Chemical control is aimed at eliminating the problem on the seed and allowing the plant to establish itself. Fungicide seed treatments approved for the control of damping-off in peas and beans include the following:

Active ingredient	*Product example*
captan + fosetyl aluminium + thiabendazole	Aliette Extra (peas)
drazoxolon	Mil-Col 30
metalaxyl + thiabendazole + thiram	Apron Combi 453 FS
thiabendazole + thiram	HY-TL, Ascot 480 FS
thiram	Tripomol 80

Wilt

(i) *Causal organism*

Fungi, *Fusarium oxysporum* f.sp. *pisi* (peas), *F. oxysporum* f.sp. *fabae* (beans).

(ii) *Host range*

Generally fungi are specific to their *forma specialis* host, but they have been isolated from a number of other hosts and will infect a

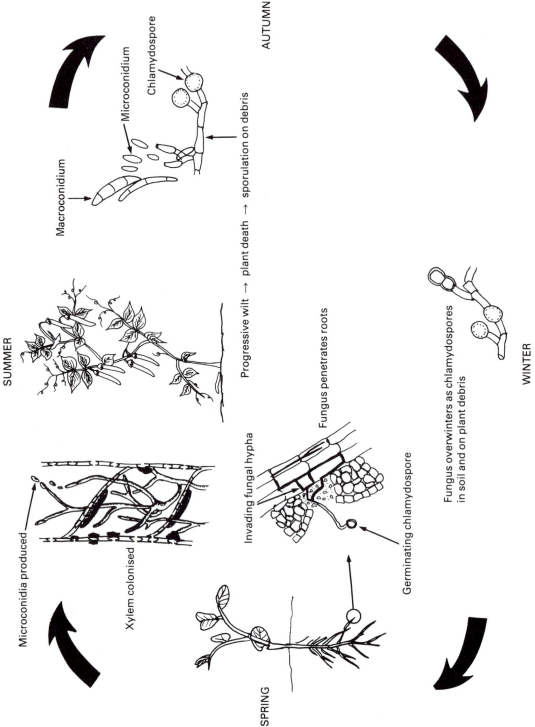

AUTUMN

Chlamydospore

Microconidium

Macroconidium

Progressive wilt → plant death → sporulation on debris

SUMMER

Fungus penetrates roots

WINTER

Microconidia produced

Xylem colonised

Invading fungal hypha

Fungus overwinters as chlamydospores
in soil and on plant debris

Germinating chlamydospore

SPRING

range of plants in artificial inoculation tests. *Fusarium oxysporum* f.sp. *pisi* exists as a number of different races.

(iii) *Symptoms*

First symptoms of *Fusarium* wilt usually become apparent during early summer. Individual plants wilt, especially during hot dry weather, and recover during the night. Later, lower leaves of the plant become chlorotic and eventually die and a permanent wilt, followed by death of the foliage, occurs from the base of the plant upwards. Plant death can occur quite quickly during dry periods around flowering time. The xylem of infected plants develops a brown-red coloration.

(iv) *Disease cycle (Figure 8.1)*

The fungi exist as chlamydospores (resting spores) in the soil and on plant debris. These are stimulated to germinate by the presence of rootlets and hyphae penetrate the rootlets. The fungus then rapidly grows towards the xylem. In the xylem, both mycelium and spores are produced. Rapid colonisation of considerable lengths of xylem may be effected by the production of microconidia which are transported in the transpiration stream. Wilting is caused both by physical blockage of xylem by the fungus and probably the production of toxins. The fungus remains confined to xylem vessels until the plant dies, when it grows out of the dead tissue and produces more chlamydospores, which are released into the soil. High temperatures, around 22 °C, are optimal for disease development, as are acid, poorly fertilised soils.

(v) *Significance*

Fusarium oxysporum f.sp. *fabae* is of minor significance on beans and has not yet been detected in Europe. *Fusarium oxysporum* f.sp *pisi* is of more significance on peas, although it has been less troublesome since the introduction of resistant varieties.

(vi) *Control*

Disease resistance is the main and most effective control method. The majority of the varieties on the NIAB Farmers' Leaflet No. 10, *Recommended Varieties of Field Peas and Field Beans* are resistant to the most common race of *F. oxysporum* f.sp. *pisi*, race 1.

Figure 8.1. Disease cycle of wilt (*Fusarium oxysporum* f.sp. *pisi* (peas), f.sp. *fabae* (beans))

Stem rot

(i) *Causal organism*

Fungi, *Sclerotinia sclerotiorum* infects peas and spring beans; *S. trifoliorum* infects winter beans.

(ii) *Host range*

Sclerotinia sclerotiorum has a very wide host range, including oilseed rape, peas and spring beans. *Sclerotinia trifoliorum* attacks clover and winter beans.

(iii) *Symptoms*

Seedlings and young plants may be affected by *Sclerotinia*, resulting in a slimy wet rot extending from the stem base to the roots. Infected areas are usually covered with white cottony fungal mycelium. Symptoms are most commonly seen in mature plants; water-soaked brown lesions first appear on stems and leaves. These areas rot quickly and may become covered with white mycelium. As the disease develops, infected plants may suddenly wilt and small dark sclerotia become visible both on the surface of decaying tissue and in the stem cavity.

(iv) *Disease cycle*

The disease cycle on peas and spring beans is similar to that on oilseed rape. There is some debate, however, as to whether damage or adhering petals is a pre-requisite for ascospore infection of peas and beans. Most infections of winter beans by *S. trifoliorum* occur in the autumn and winter and not at flowering.

(v) *Significance*

The disease can be found in most areas where peas and beans are grown, but severe epidemics are uncommon. Individual affected plants usually yield very little and contamination of the harvested crop, particularly vined peas (peas harvested green for human consumption) by sclerotia, can lead to rejection.

(vi) *Control*

Cultural control. Contaminated straw should be burned and groundkeepers eliminated. Good crop rotation may also reduce the disease, but many arable crops are susceptible.

Downy mildew

(i) *Causal organism*
Fungus, *Peronospora viciae*.

(ii) *Host range*
The most important hosts are peas and beans.

(iii) *Symptoms*
Young plants may be systemically colonised by *P. viciae*. Plants are stunted and chlorotic and the undersides of leaflets become covered with dense grey-white fungal hyphae. Systemically colonised plants usually die. Disease in mature plants results in chlorotic patches on upper leaf surfaces with a corresponding grey-white fungal growth on the underside of the leaf. Pods and seeds can also become contaminated with downy mildew. Pods develop a yellow blotchiness externally, leading to white fungal growth inside the pod and discoloration of seeds.

(iv) *Disease cycle*
The main source of overwintering inoculum is oospores in contaminated debris. The oospores can survive in the soil and seed may also be contaminated. Oospores in the surface layers of soil or debris germinate and infect plants, resulting in systemic colonisation, or produce sporangia which initiate infections on aerial parts of the plant. Further sporangia are produced on the undersides of the leaves and are dispersed by rain and wind. On susceptible host tissue sporangia germinate directly leading to penetration of the leaf surface. Spore germination is optimal at temperatures between 4 and 8 °C and long periods of high humidity are also required.

(v) *Significance*
The disease is more significant on peas than on beans. In cool wet seasons, plant populations can be reduced by systemic colonisation and yield losses of over 50% have been reported in early-sown vining peas. Quality of vining peas in particular is reduced by downy mildew. On beans the disease should not generally be a serious problem, although some crops of spring beans have been severely affected in the UK in recent years.

(vi) *Control*

Cultural control includes good rotations, allowing at least four years between pea crops in badly infested land. Debris should also be disposed of properly, preferably by burning.

Disease resistance ratings are given for field peas in NIAB Farmers'
Leaflet No. 10, *Recommended Varieties of Field Peas and Field
Beans*.

Chemical control methods are available. Seed treatments, containing the systemic chemicals fosetyl-aluminium or metalaxyl, are used
to eradicate seed surface contamination and give early season
protection from further attacks in pea crops in particular. Foliar
sprays may also be applied to peas. Products approved for control of
downy mildew include:

Active ingredient	*Product example*
captan + fosetyl-aluminium + thiabendazole	Aliette Extra (seed treatment: peas and beans)
carbendazim + chlorothalonil	Bravocarb (peas)
chlorothalonil	Bravo 500 (peas)
chlorothalonil + metalaxyl	Folio 575 FW (beans)
fosetyl-aluminium	Aliette
mancozeb + metalaxyl	Fubol 58 WP (beans)
metalaxyl + thiabendazole + thiram	Apron Combi 453 FS (seed treatment)

Powdery mildew (peas)

(i) *Causal organism*
Fungus, *Erysiphe pisi* (*Erysiphe polygoni*).

(ii) *Host range*
Peas, lucerne, vetches, chick pea, pigeon pea and lentil. Beans are
not usually host species.

(iii) *Symptoms*
All aerial parts of the plant can be affected. Usually, the upper
surface of the leaf suffers first as small discrete grey-white fungal
pustules develop. Larger areas may then be colonised and patches
of chlorotic and necrotic tissue often occur. Black cleistothecia can
occasionally be seen in older mildew pustules.

(iv) *Disease cycle*
The significance of cleistothecia in overwintering is not well
understood. Generally the main source of overwintering inoculum
is resting mycelium in early-sown autumn crops, groundkeepers and
green debris. The pathogen may also overwinter on seeds. Light,

aerially dispersed powdery mildew conidia result in secondary infections and the disease seems to be most severe after warm dry days and heavy dews in crops in sheltered locations.

(v) *Significance*
Nationally, the disease is of little significance. However, it can be troublesome in individual crops.

(vi) *Control*
Cultural control involving elimination of groundkeepers and diseased haulm tissue may be useful.

Disease resistance is being introduced into new pea varieties grown in the USA.

Chemical control using foliar sprays of sulphur will reduce the disease.

Rust

(i) *Causal organism*
Fungi, *Uromyces pisi* (peas) and *U. fabae* (beans and peas).

(ii) *Host range*
Peas, beans and a few other plant species including lentils.

(iii) *Symptoms*
The disease is often seen late in the growing season on leaves of peas and beans. Red-brown small rust pustules often surrounded by chlorotic haloes are randomly scattered over leaf surfaces. Early infections can lead to defoliation of plants.

(iv) *Disease cycle*
The fungus overwinters primarily as uredospores or resting mycelium on living plant material, provided by volunteer beans, old green bean debris and autumn-sown crops. High humidity encourages the disease.

(v) *Significance*
Uromyces fabae is the more important pathogen and it is more widespread on beans than peas. However, most late-season attacks reduce yields only slightly, although yield losses of up to 30% have

been reported in artificially infected trials. The disease may be more severe in soils with a low potassium content.

(vi) *Control*

Cultural control. Elimination of volunteers and contaminated debris will reduce disease.

Chemical control can be undertaken by foliar sprays, which are particularly useful to halt early build-up of disease. Fungicides approved for application to beans include fenpropimorph (Corbel).

Ascochyta **leaf stem and pod spot**

(i) *Causal organism*
Fungi, *Ascochyta pisi* (peas), *A. fabae* (beans).

(ii) *Host range*
Mainly peas and beans.

(iii) *Symptoms (Figure 8.2)*
Initially, small brown circular lesions, which may be confused with chocolate spot, develop on leaves. However, as the lesions develop, they become slightly sunken and have pale grey centres with black pycnidia. They also become more irregular in shape and it is common for lesions to coalesce and 'run' to the edges of leaves. On stems, elongated spreading dark brown sunken streaks develop, containing black pycnidia. On pods, early infection causes necrosis and may result in abortion. Lesions which develop later consist of circular areas of dark brown sunken tissue, the pale centres containing many black pycnidia. Should the fungus penetrate the pod, white fungal growth can be seen inside and seeds become stained.

(iv) *Disease cycle*
Primary infections develop from inoculum on seed and crop debris, although *A. pisi* can survive saprophytically in the soil for short periods. Spores released from pycnidia are splash-dispersed horizontally from plant to plant and vertically up infected plants. The disease develops best during periods of high humidity and at temperatures of about 20 °C.

(v) *Significance*
Both yield and quality of crops can be significantly reduced by severe attacks of *Ascochyta*. Yield losses of up to 30% have been

reported in beans. There is a maximum tolerance of 0.2% infection in field bean crops grown for seed in the UK. Disease levels in excess of this result in rejection of the crop for seed.

(vi) *Control*
Cultural control includes the use of certified seed. Debris should also be burned if possible.

Figure 8.2. *Aschochyta* pod spot of beans caused by *Ascochyta fabae*. (A. J. Biddle, PGRO.)

Chemical control. Seed treatments are useful in eliminating contamination of seed and reducing early attacks of the disease especially in peas. Foliar sprays are particularly useful in crops grown for quality markets, vining peas for example, or in field bean crops grown for seed. Products approved for the control of *Ascochyta* include:

Active ingredient	*Product example*
benomyl	Benlate (seed treatment)
captan + fosetyl-aluminium + thiabendazole	Aliette Extra (seed treatment, peas)
carbendazim + chlorothalonil	Bravocarb (peas)
chlorothalonil	Bravo 500 (peas)
metalaxyl + thiabendazole + thiram	Apron Combi 453 FS (seed treatment)
thiabendazole + thiram	HY-TL (seed treatment, peas)

Mycosphaerella leaf stem and pod spot

(i) *Causal organism*
Fungus, *Mycosphaerella pinodes* (*Ascochyta pinodes*).

(ii) *Host range*
Mainly peas, although beans may be affected.

(iii) *Symptoms (Figure 8.3)*
Post-emergence damping-off may be a consequence of growing seed contaminated by *M. pinodes*. Lesions on leaves initially consist of small 'tea-leaf' shaped discrete dark brown or purple-brown patches. Lesions may enlarge and assume a concentric ring pattern. Infection of stems results in small brown streaks at first, which may later coalesce, girdling the stem and giving it a blue-black colour. Lesions on pods develop in a similar way to lesions on leaves.

(iv) *Disease cycle*
Inoculum overwinters on seed and crop debris. Ascospores released from pseudothecia, which are produced mainly on diseased stems and pods, usually initiate spring infections. Pycnidia are also produced, which release splash-dispersed asexual conidia. High relative humidity and temperatures of about 20 °C favour disease development.

(v) *Significance*
The disease is widely distributed in both temperate and sub-tropical areas and losses can be significant as a result of severe disease attacks, particularly if all phases (damping-off, foot rot and leaf stem and pod spot) develop.

(vi) *Control*
As *Ascochyta*.

Figure 8.3. Initial symptoms of *Mycosphaerella* leaf spot of peas caused by *Mycosphaerella pinodes*. (J. Thomas, NIAB.)

Grey mould (peas)

(i) *Causal organism*
Fungus, *Botrytis cinerea* (*Botryotinia fuckeliana*).

(ii) *Host range*
Ubiquitous facultative parasite attacking almost any senescing or decaying tissue.

(iii) *Symptoms*
All aerial parts of the plant may be affected. Irregular, water-soaked lesions develop, especially around damage points or rotting petals. Tissue may rot completely resulting in a characteristic bloom of grey mould.

(iv) *Disease cycle*
As in oilseed rape. Wet humid weather is essential for disease development.

(v) *Significance*
The disease is widespread, but generally occurs at low levels unless there is a prolonged wet period in summer. If harvesting is delayed because of rain, *Botrytis* may become a problem. Pods may be more susceptible to grey mould in soil deficient in potassium.

(vi) *Control*
Cultural control methods are useful. They include use of optimum seed rates and avoidance of oversheltered places to reduce humidity within the crop. Any potassium deficiency in the soil should be corrected.

Chemical control of the disease can be undertaken by foliar sprays. Fungicides approved for control include:

Active ingredient	Product example
benomyl	Benlate
carbendazim + chlorothalonil	Bravocarb
chlorothalonil	Bravo 500
iprodione	Rovral Flo
vinclozolin	Ronilan

Chocolate spot (beans)

(i) *Causal organism*
Fungi, *Botrytis fabae* and *B. cinerea.*

(ii) *Host range*
Botrytis fabae is restricted to beans, *B. cinerea* occurs on numerous hosts.

(iii) *Symptoms (Figure 8.4)*
Symptoms of chocolate spot are divided into two distinct phases; non-aggressive and aggressive. Non-aggressive chocolate spot, which can be caused by either *Botrytis* species, consists of small discrete red-brown lesions, usually on leaves although stems and pods can also be affected. Lesions are generally evenly distributed over tissue surfaces giving a 'peppered' appearance. Aggressive chocolate spot lesions are much larger and darker in colour. Lesions coalesce and large areas of tissue may die. Leaves can be totally destroyed and may be shed. Infected flowers and pods may abort. On stems, dark brown streaks and lodging may occur. This much more significant phase of the disease is generally attributed to *B. fabae*, although *B. cinerea* can be induced to form aggressive chocolate spot in artificial inoculation experiments by the addition of a small amount of pollen or pollen extract.

(iv) *Disease cycle (Figure 8.5)*
Both species of *Botrytis* can form sclerotia. However, they are a more significant means of overwintering in *B. fabae* than *B. cinerea*

Figure 8.4. Non-aggressive chocolate spot of beans caused by either *Botrytis cinerea* or *B. fabae*. (J. Thomas, NIAB.)

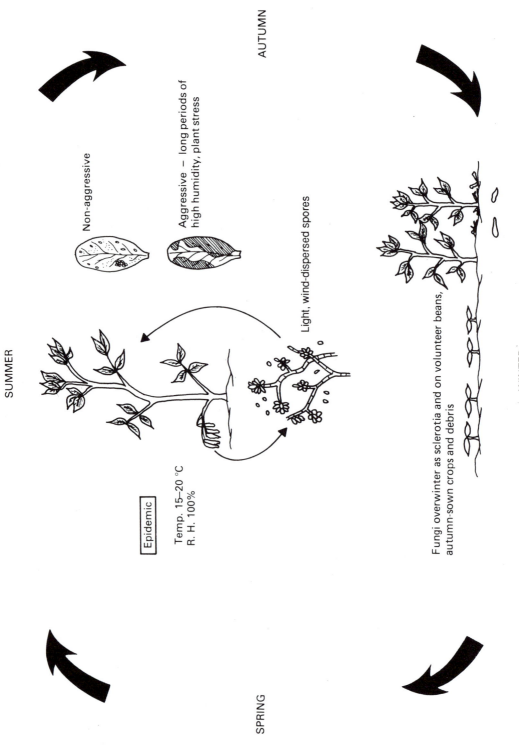

AUTUMN

Non-aggressive

Aggressive – long periods of
high humidity, plant stress

Light, wind-dispersed spores

SUMMER

Epidemic

Temp. 15–20 °C
R. H. 100%

Fungi overwinter as sclerotia and on volunteer beans,
autumn-sown crops and debris

WINTER

SPRING

because of the restricted host range of the former. Volunteer beans, autumn-sown crops and debris are other possible ways by which the fungi overwinter. During mild humid weather (15–20 °C and 100% R.H. optimal), wind-dispersed conidia are produced which initiate infections. The development of the aggressive phase seems to be dependent upon long periods of high humidity. Acid soils, mineral deficiency, waterlogging and high plant density also encourage the aggressive form of the disease.

(v) *Significance*
The significance of the disease is related to severity and timing of attack. Non-aggressive chocolate spot, although dramatic in appearance does not usually affect yield significantly. Aggressive chocolate spot occurring relatively early in the season can reduce yield by 50%.

(vi) *Control*
Cultural control of the disease is aimed at reducing the chance of the aggressive phase developing. Reducing humidity in crops by optimum seed rates and avoiding oversheltered areas will help. Reducing crop stresses by improving drainage and using optimum fertiliser rates also reduces disease.

Chemical control is widely practised against the disease. Foliar-applied fungicides may be used at regular intervals during the growing season, depending on levels of disease and weather conditions. Fungicides approved for control of chocolate spot include:

Active ingredient	*Product example*
benomyl	Benlate
bromophos + captan +thiabendazole	Bromotex T (seed treatment)
carbendazim	Bavistin, Derosal Liquid
carbendazim + chlorothalonil	Bravocarb
chlorothalonil	Bravo 500
iprodione	Rovral Flo
iprodione + thiophanate-methyl	Compass
thiabendazole + thiram	Hy-Vic (seed treatment)
thiophanate-methyl	Cercobin Liquid
vinclozolin	Ronilan FL

Figure 8.5. Disease cycle of chocolate spot (*Botrytis fabae*, *B. cinerea*).

Note that resistance to the MBC fungicides (benomyl, carbendazim and thiophanate-methyl) and the dicarboximides (iprodione and vinclozolin) has been detected in *B. fabae* and *B. cinerea*.

Bacterial blight (peas)

(i) *Causal organism*
Bacterium, *Pseudomonas syringae* pv. *pisi*.

(ii) *Host range*
Peas. Five races of *P. syringae* pv. *pisi* have been reported.

(iii) *Symptoms*
Lesions initially consist of small round dark green water-soaked spots on leaves and pods. Later, lesions enlarge, become necrotic and papery and may be angular in appearance. The lesions are often limited by veins. In wet weather, a creamy bacterial ooze can be produced in lesions. On pods, lesions tend to be sunken and green-brown in colour. Severe attacks lead to infection of seeds; they become covered with a bacterial slime and may also be stained.

(iv) *Disease cycle*
The pathogen overwinters mainly on seed, but volunteers and autumn-sown crops may also harbour it. Spread of the pathogen occurs mainly by rain-splash. Contact spread, and spread by contaminated machinery and insects have also been implicated. Cool humid summer weather encourages bacterial ooze production.

(v) *Significance*
The disease occurs in most pea-growing areas of the world, but is particularly common in the USA. The first record of its occurrence in the UK was in 1986. Should the disease become established in the UK, losses of 10% have been predicted in susceptible varieties in suitable environmental conditions.

(vi) *Control*
Bacterial blight of peas is a notifiable disease in the UK. Crops contaminated with the pathogen must be destroyed. The use of resistant varieties and pathogen-free seed are the main methods by which the disease is controlled in the USA.

Virus diseases
Thirty-four viruses are listed by Cockbain (1983) as occurring naturally in field beans. Five of the major pea virus diseases are discussed by Hagedorn (1985) (pea mosaic, pea enation mosaic, pea top yellows, pea seed-borne mosaic and pea streak). Hagedorn

(1984) lists 11 viruses which naturally infect peas in the USA. The main vectors of virus diseases are aphids, beetles and nematodes. Few virus infections appear to result in significant yield losses. The main methods of control include use of resistant varieties and control of aphid vectors. It may be more desirable to control aphids because of the damage they do to peas and beans, irrespective of virus transmission. Products approved for control are given in the *UK Pesticide Guide* (Ivens, 1989).

Further reading

Cockbain, A. J. (1983). Viruses and virus-like diseases of *Vicia faba* L. In *The Faba Bean* (*Vicia faba* L.), ed. P. D. Hebblethwaite, pp. 421–62. London: Butterworths.

Dixon, G. R. (1981). *Vegetable Crop Diseases*. London: Macmillan.

Gane, A. J., Biddle, A. J., Knott, C. M. & Eagle, D. J. (1984). *Pea Growing Handbook*. Processors & Growers Research Organisation, Peterborough.

Gaunt, R. E. (1983). Shoot diseases caused by fungal pathogens. In *The Faba Bean* (*Vicia faba* L.), ed. P. D. Hebblethwaite, pp. 463–92. London: Butterworths.

Hagedorn, D. J. (ed.) (1984). *Compendium of Pea Diseases*. St Paul: American Phytopathological Society.

Hagedorn, D. J. (1985). Diseases of peas: their importance and opportunities for breeding for disease resistance. In *The Pea Crop*, ed. P. D. Hebblethwaite, M. C. Heath & T. C. K. Dawkins, pp. 205–13. London: Butterworths.

Harrison, J. G. (1988). The biology of *Botrytis* spp. on *Vicia* beans and chocolate spot disease – a review. *Plant Pathology*, **37**, 168–201.

National Institute of Agricultural Botany (1989). Farmers' Leaflet No. 10, *Recommended Varieties of Field Peas and Field Beans*. Cambridge: NIAB publication.

Salt, G. A. (1983). Root diseases of *Vicia faba* L. In *The Faba Bean* (*Vicia faba* L.) ed. P. D. Hebblethwaite, pp. 393–419. London: Butterworths.

9 Diseases of potatoes

Potatoes are the world's most important non-cereal crop. The area of potatoes grown in the UK has halved during the last 25 years and currently (1988) stands at approximately 160 000 ha (England and Wales). Potato production in the UK is organised on a quota system by the Potato Marketing Board. Potato crops are highly productive with average yields of main-crop (late maturing) and second early potatoes between 1982 and 1986 of 36.3 t/ha. Although there are considerable price fluctuations in the potato market, the crop can be very profitable.

Diseases of potatoes include arguably the most historically significant crop disease, late blight, which is still the most important potato disease. An increasing emphasis on the cosmetic appearance of potatoes has recently brought hitherto non-significant diseases into prominence.

Disease may affect either the haulm (stems and leaves) or the tuber or both. Some diseases are of particular significance in stored potatoes. Many diseases can be carried on and in 'seed' tubers. There is therefore much emphasis on the production and use of high quality, disease-free seed.

Potato blight (late blight)

(i) *Causal organism*
Fungus, *Phytophthora infestans*.

(ii) *Host range*
Members of the Solanaceae, especially potato and tomato. Pathovars or races of *P. infestans* exist.

(iii) *Symptoms (Figure 9.1)*

(a) *on haulm*. The first signs of potato blight are circular or irregular water-soaked patches, often at the tips or edges of lower leaves. Large brown dead areas of leaves may quickly develop and a zone of white downy fungal growth may occur at the edge of the lesion on the underside of the leaf. The disease can progress very quickly through the foliage during wet weather and reduce the green tissue to a brown, limp stinking mess. During periods of dry weather, production of fresh lesions is inhibited.

(b) *on tuber*. The first signs of blight in potato tubers are unobtrusive brown or black irregular blotches. Tubers may also

Figure 9.1. Late blight of potatoes caused by *Phytophthora infestans* (i) on leaves, (ii) on tubers. (University of Wisconsin.)

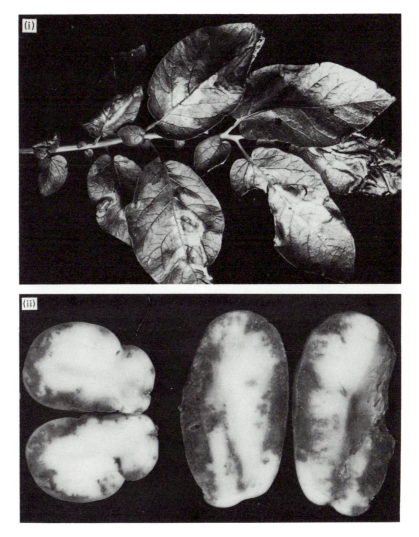

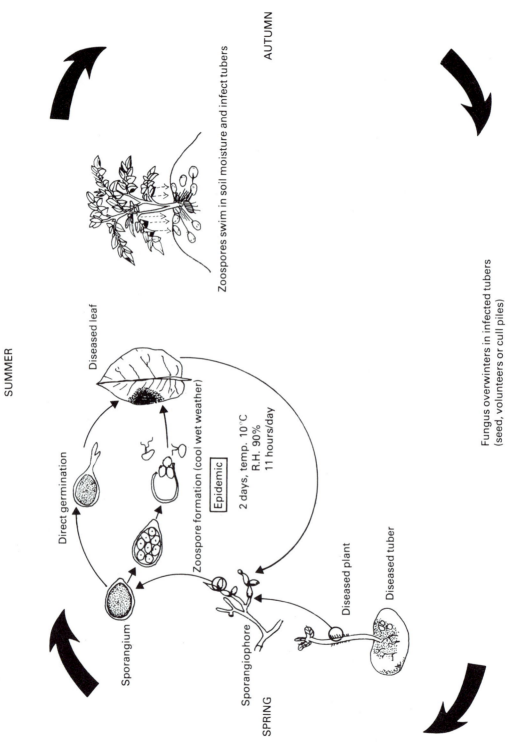

AUTUMN

Zoospores swim in soil moisture and infect tubers

SUMMER

Diseased leaf

Direct germination

Sporangium

Zoospore formation (cool wet weather)

Epidemic

2 days, temp. 10°C
R.H. 90%
11 hours/day

Sporangiophore

SPRING

Diseased plant

Diseased tuber

Fungus overwinters in infected tubers
(seed, volunteers or cull piles)

WINTER

appear 'marbled' in patches where internal streaks of brown tissue can be seen through the skin. Blight symptoms are best seen when tubers are cut open. The first symptom is a brown discolouration of the outside layers of tuber tissue. This can quickly progress inwards, resulting in large areas of firm but brown rotten tissue. Blight can progress in infected tubers in potato stores but it does not usually spread to healthy tubers. Blighted tubers are frequently colonised by secondary bacterial pathogens, particularly in poorly ventilated, warm humid potato stores. Tubers are then very quickly reduced to a semi-liquid state.

A descriptive key for assessment of potato blight is given in Chapter 3.

(iv) *Disease cycle (Figure 9.2)*

Phytophthora infestans overwinters as mycelium in contaminated potato tubers. These may be in cull (discarded) piles of potatoes, groundkeepers left from a previous crop, or seed stocks. Cull piles are probably the most important source of initial inoculum in the UK. Diseased tubers give rise to diseased haulm tissue. During suitable weather conditions (see Control: Forecasting disease), sporangiophores are produced which bear many lemon-shaped sporangia. These can be dispersed relatively short distances by rain-splash or longer distances on air-currents. Sporangia require over 90% relative humidity for germination. They may germinate directly, penetrate healthy tissue and, within a matter of 4 days during optimum weather conditions, produce more sporangia. However, particularly during slightly cooler, wet weather, each sporangium releases 8–12 motile biflagellate zoospores which encyst and then penetrate tissue directly or, occasionally, via stomata. These zoospores are important in the cycle of contamination of tubers. During wet weather, tubers near the soil surface may be infected by zoospores which swim in soil moisture. Again zoospores encyst prior to penetrating the tuber via wounds or lenticels. Tubers may also become infected if they are exposed to airborne sporangia from haulm tissue or to sporangia and zoospores in the top layer of soil during harvesting. A further interesting feature of the life cycle of *P. infestans* concern the possibility of sexual reproduction resulting in oospores. Until recently, only one of the two mating types needed for sexual reproduction could be found in the UK. However, the complementary mating type, previously confined to Mexico, has been identified in blight-affected crops in the UK (Tantius *et al.*, 1986).

Figure 9.2. Disease cycle of late blight (*Phytophthora infestans*).

(v) *Significance*

Potato blight is undoubtedly the most important disease of potatoes. The disease altered the course of human history during the 19th century. Potatoes had become the staple diet of poor Irish people in the first part of the 19th century. They were eaten in such large quantities that they provided an adequate diet. During 1845, potato blight swept across Europe and into Ireland resulting in 25% yield loss. The disease was even more severe during 1846 when about 80% of the crop was lost. One quarter of the Irish population of 8 million died through starvation and a further 1 million emigrated, mostly to North America, during the subsequent decade. In England, the famine led to the repeal of the Corn Laws and the fall of the British government. Obviously the disease is not as significant today, mainly because of effective methods of control. The precise relationship between yield potential of the crop, damage and crop loss has not been established (Rotem, Bashi & Kranz, 1983), but it is evident that early attacks of blight, if not controlled, reduce yields dramatically. Work by Large in the 1950s suggested that tuber development stopped when 75% of the foliage was destroyed by blight. The earlier in the season this level of blight was reached, the greater the yield loss incurred. For example, if 75% foliage blight occurred at the end of July, Large calculated that there would be approximately 50% yield loss. However, if the threshold level was reached during mid-September, there would be only about 4% yield loss (Large, 1952, 1958).

Should sexual reproduction in *P. infestans* occur in the UK, new strains could develop by hybridisation between the sexually compatible mating types.

(vi) *Control*

Cultural control methods are important in blight control. Rotations should be planned to avoid planting early and late maturing (main) potato crops in adjacent fields. The former are less frequently treated with fungicides to control blight, and the disease may develop to a limited extent, thus providing inoculum for infection of nearby later-sown main crops. Cull piles of waste potatoes should be disposed of properly and not allowed to sprout. Covering with black plastic sheets or spraying with herbicides will kill sprouts. It is desirable to reduce the chance of introducing blight in seed tubers. Certified seed should have very low levels of blight, but it is still advisable to allow seed tubers to sprout, and reject tubers which do not sprout. Planting even a few contaminated seed tubers may produce a focus of disease. Groundkeepers should be eliminated, and good deep ridges made in the field to reduce the potential for tuber infection.

Disease resistance exists in potato varieties, but is not sufficient on its own to adequately control disease. In the UK, NIAB Farmers' Leaflet No. 3, *Recommended Varieties of Potatoes*, gives resistance ratings to both foliage and tuber blight. Some varieties, most notably the maincrop variety King Edward, are highly susceptible to potato blight and should be avoided if possible in areas where blight is a severe problem.

Chemical control methods are heavily relied upon to give protection from potato blight, but some problems have recently developed in control of blight by fungicides. The value of the crop, coupled with the losses which can occur as a result of blight, have led to common use of intensive fungicide spray programmes. These programmes vary, but there are basically two approaches. The first involves the use of protectant fungicides, e.g. dithiocarbamates. These must be applied before blight is seen in the crop and repeated applications are necessary to allow for weathering and to protect new growth. Most product recommendations suggest a first application as plants meet in the row and subsequently at 10–14 day intervals. There are at present no limitations on number of applications of protectant fungicides. However, products containing fentin hydroxide and fentin acetate are currently under review regarding their mammalian toxicity. The aim of a protectant spray programme would ideally be to delay blight indefinitely. However, in most seasons, protectant programmes merely delay the time at which 75% foliage affected is reached.

The second approach to chemical control of blight involves the use of systemic fungicides. Since the early 1980s, fungicides containing phenylamides as active ingredients (e.g. metalaxyl) have been used as highly effective blight sprays. They offer much better control of blight in that they are resistant to weathering and can protect new growth. However, largely as a result of the development of resistant isolates of *P. infestans*, there are restrictions on the use of systemic blight fungicides. Use of metalaxyl alone in the Netherlands and Republic of Ireland failed to control blight in 1980. This occurred after only 2 years' usage. As a consequence of this, phenylamides are sold in the UK as pre-pack mixed formulations containing a protectant dithiocarbamate fungicide. This has prolonged the useful life of the phenylamides, but there were suggestions of reduced control in the field during the 1987 season. The first application of systemic fungicides should occur before blight is seen, i.e. when plants meet along the rows. Depending on the risk of infection, repeated applications are recommended at intervals of 10–21 days, with a maximum of five applications per season. Most growers now use a combination of protectant and systemic sprays. It is common for systemics to be used during the first part of the season and protectants towards the end. Products containing fentin hydroxide or fentin acetate are particularly popular as end-of-season sprays, partly because they give additional protection to the tuber against blight. There is, as already stated, some doubt as to the future of these chemicals in this particular role. Indeed, fungicidal control of potato blight as a whole is becoming increasingly problematical. Fungicides approved for control of blight at present include the following:

Active ingredient	Product example
benalaxyl + mancozeb	Galben M
chlorothalonil	Bravo 500, Bombardier
copper hydroxide	Chiltern Kocide 101
copper oxychloride	Cuprokylt
copper sulphate	Comac Bordeaux Plus
copper sulphate + cufraneb	Comac macuprax
copper sulphate + sulphur	Stoller
cymoxanil + mancozeb	Fytospore
fentin acetate + maneb	Brestan 60
fentin hydroxide	Du-Ter 50
fentin hydroxide + maneb + zinc	Chiltern Tinman
ferbam + maneb + zineb	Trimanzone
mancozeb	Dithane 945, Penncozeb
mancozeb + metalaxyl	Fubol 75 WP
mancozeb + oxadixyl	Recoil
maneb	Manzate
maneb + zinc	Manex
maneb + zinc oxide	Mazin
manganese zinc ethylenebisdithiocarbamate complex	Trithac
manganese zinc ethylenebisdithio-carbamate + ofurace	Patafol Plus

| zineb | Zineb |
| zineb-thiuram ethylene disulphide complex | Polyram |

Forecasting disease: Blight forecasting is perhaps the most widely used plant disease forecasting system in the UK. *Phytophthora infestans* development is very weather-dependent and its precise temperature and humidity requirements have been deduced after many years' work. These have been incorporated into 'Smith' or blight infection periods in the UK. A Smith Period is defined as two consecutive days (ending at 0900 hours) when the temperature has not been less than 10 °C and the relative humidity has been above 90% for at least 11 hours of each day.

Notifications of Smith Periods are issued when deemed relevant, having regard to the date and stage of development of the crop. In the UK, the ADAS are responsible for communicating this information on a regional basis through bulletins, farming press, television, radio, telephone information services and on 'Prestel'. In the USA, a computerised forecast for potato blight has been developed called BLITECAST, again based on temperature and relative humidity thresholds. The programme is available on microcomputer which can be coupled to a weather data logger and the two situated in a potato field. In areas where potato blight is not annually a severe problem, disease forecasts can be used effectively to determine timing of fungicide application. However, in many areas of the UK, farmers cannot afford to take the risk of reducing spray applications and a routine ('insurance') spray programme is adopted.

Integrated control of potato blight is becoming increasingly important, particularly with the growing problems associated with fungicide usage. Perhaps there has been an over-reliance on fungicides and neglect of other methods of disease control, which may by necessity change in the future.

Early blight (target spot)

(i) *Causal organism*
Fungus, *Alternaria solani*.

(ii) *Host range*
Solanaceae, including tomato and potato, and some *Brassica* species.

(iii) *Symptoms*

In the UK, symptoms consist of dark brown circular to oval leaf spots which frequently contain concentric brown circles. Mature lesions may be limited by leaf veins. Symptoms may be confused with late blight, but normally *Alternaria* occurs earlier in the season and there is no downy fungal growth associated with lesions. In warmer countries, *A. solani* can rot tubers.

(iv) *Disease cycle*

The fungus overwinters as conidia, mycelium or possibly chlamydospores on alternative hosts, diseased crop debris or seed potatoes. The large club-shaped conidia are wind and water-dispersed in the crop. Dry windy conditions favour conidial release, but moisture is essential for sporulation and spore germination. Optimum temperatures for disease development are normally above 20 °C.

(v) *Significance*

The disease is not of economic significance in the UK.

(vi) *Control*

Unnecessary in the UK. Fungicides used for the control of late blight would probably be effective against early blight.

Stem canker and black scurf

(i) *Causal organism*

Fungus, *Rhizoctonia solani* (*Thanatephorus cucumeris*).

(ii) *Host range*

A wide range of crops including many *Brassica* species, beans sugar beet, tomatoes and potatoes. There is some differentiation of the fungus into strains.

(iii) *Symptoms (Figure 9.3)*

(a) *on haulm (stem canker)*. The fungus can kill sprouts, resulting in gaps or delayed emergence. Symptoms first appear on underground stems as brown cankers which may girdle the stems that emerge. Severe cankers can result in the formation of aerial tubers and cause rolling and wilting of foliage. Another characteristic symptom of stem canker is the formation of a white powdery collar, again

Figure 9.3. Stem canker
and black scurf of potatoes
caused by *Rhizoctonia
solani*. (i) Stem canker
(M. C. Shurtleff); (ii)
black scurf on tuber (Plant
Breeding International,
Cambridge).

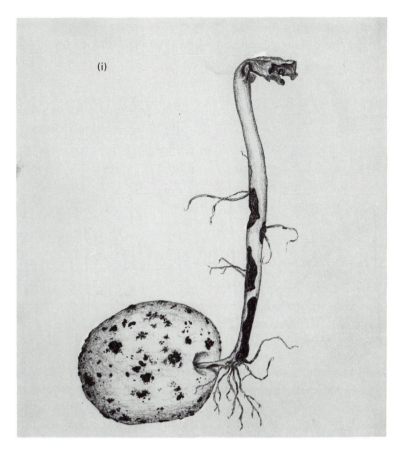

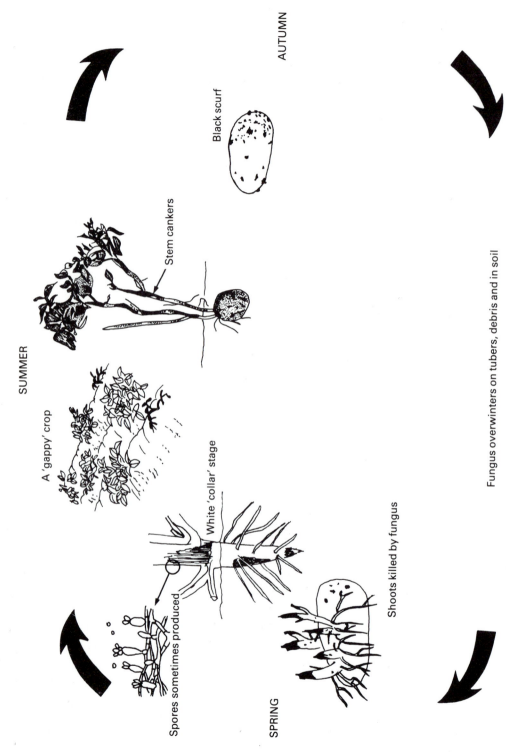

AUTUMN

Black scurf

Stem cankers

SUMMER

A 'gappy' crop

Fungus overwinters on tubers, debris and in soil

WINTER

White 'collar' stage

Spores sometimes produced

Shoots killed by fungus

SPRING

Table 9.1. *Summary of differences between stem canker*
(Rhizoctonia solani) *and blackleg* (Erwinia carotovora *subsp.*
atroseptica)

Stem canker	Blackleg
Brown canker on haulm with clearly defined edge	Dark brown-black soft lesion with diffuse edge on haulm
White powdery collar girdling stem just above ground level may develop	—
Aerial tubers may form	
—	Dark brown streaks high up stem may occur
—	Brown xylem
—	Rotten seed tuber

girdling the stem, just above ground level. This is seen most
frequently during humid early summer weather. A summary of the
main differences between stem canker and blackleg, a disease
which produces similar symptoms, is given in Table 9.1.

(b) *on tubers (black scurf)*. Black or dark brown sclerotia develop
on the surface of mature tubers. These may occur individually or
aggregate to form large patches. They are loosely attached to the
skin and can easily be scratched off.

(iv) *Disease cycle (Figure 9.4)*
The fungus survives the winter as sclerotia or mycelium on seed
tubers, in the soil or on debris. Mycelium infects potato stems or
emerging sprouts in spring and the disease appears to be most
severe in dry light soils, particularly when conditions are cold at the
time of planting. Spores may be produced on the white 'collar' stage
but the disease usually spreads underground from plant to plant by
mycelial growth. Tubers may became infected at any time during
growth and sclerotia are formed towards the end of the season.

(v) *Significance*
A potato crop suffering from stem canker shows a good ability to
compensate for the disease. However, there may still be about 10%
yield loss of tubers in a severely affected crop. Severe black scurf on
the tubers may be much more significant. The 'cosmetic' damage
results in reduced marketability or perhaps total rejection by some

Figure 9.4. Disease cycle
of stem canker and black
scurf (*Rhizoctonia solani*).

buyers involved in the sale of high quality pre-packed potatoes. The disease tends to be seasonal and is only a problem in certain soils.

(vi) *Control*

Cultural control involves, amongst other methods, decreasing the time between planting and emergence. Physiologically aged seed tubers or those with good strong sprouts are less at risk from stem canker. Cold conditions should be avoided at planting and good crop rotations may also help to reduce the disease.

Chemical control methods are becoming more popular for the control of the disease. Fungicides are generally applied to the seed potatoes just before planting, or in the planting furrow. However, treating seed tubers at harvest or during loading into store may also be effective at reducing disease development in store. Fungicides approved for control of black scurf or stem canker include:

Active ingredient	*Product example*
iodophor + thiabendazole	Byatran, Tubazole
pencycuron	Monceren DS
quintozene	Tubergran
tolclofos-methyl	Rizolex

It is unlikely that routine treatment of all seed potatoes in all circumstances is worthwhile. Fungicides should be used only on crops grown for a high quality market or on farms with a history of severe stem canker.

Blackleg and soft rot

(i) *Causal organism*

Bacteria, *Erwinia carotovora* subsp. *atroseptica* causes both blackleg and soft rot in the UK. *Erwinia carotovora* subsp. *carotovora* also causes soft rot.

(ii) *Host range*

The only important host is potatoes.

(iii) *Symptoms (Figure 4.5, p. 73)*

(a) *on haulm*. One of the most severe symptoms is gaps in the crop as a result of young shoots being killed before emergence. Normally, however, symptoms in mature crops are very dark brown or black lesions above the point of attachment to the seed tuber, sometimes extending above soil level. Leaves may roll and become chlorotic. During wet weather the relatively restricted lesions may enlarge rapidly, resulting in dark soft rotten stems and eventually dead plants. Xylem browning is common in blackleg as is rotting of the seed tuber (Table 9.1). Should the disease occur late in the season, bacteria can be splashed high up the stem, resulting in dark brown streaks.

(b) *on tuber*. Bacteria spread from rotting stems along stolons and give rise to a soft or wet rot which develops in the tuber from the heel end. During wet conditions, either in the field during a rainy summer or in a poorly ventilated, warm humid store, the tuber rot will develop quickly, causing total distintegration of tissue. Alternatively, it is possible in dry conditions for rot to be confined to a relatively small area around the heel end of the tuber.

(iv) *Disease cycle (Figure 9.5)*
The bacterium overwinters mainly deep in the lenticels of seed tubers. Development of the disease on the haulm is dependent upon factors including the amount of inoculum on seed, soil temperatures, humidity and presence of wounds. The bacterium appears to remain latent on the seed tuber until the precise as yet unknown, combination of factors required for activation occurs. The disease develops sporadically during the season, although there is some evidence that rotting tissue may spread bacteria to nearby neighbours. Pathogens on individual tubers have different latent periods resulting in haulm infections at various times during the season. Bacteria colonise shoots and, if vascular tissue is colonised, they invade stems and daughter tubers. Contaminated tubers, especially if damaged, can be the source of devastating soft rot in poor potato stores.

(v) *Significance*
Yield losses as a result of the disease on the haulm are generally not high. However, soft rot losses in store are much more important and as much as 5% of national crops in north-west Europe may be lost annually as a result of soft rot in store. The disease is also of considerable significance in potato seed production. Maximum tolerances are set for potatoes in the various seed categories. However, it is often difficult to detect small amounts of blackleg in a growing crop and the bacteria can spread from rotting seed to

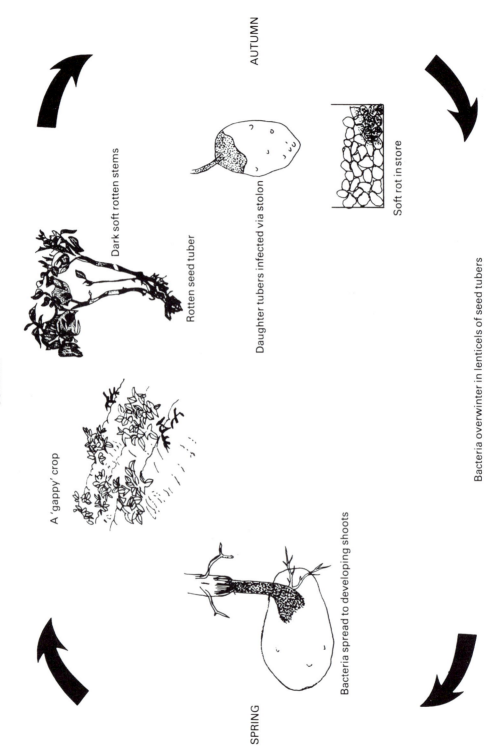

AUTUMN

Dark soft rotten stems

Rotten seed tuber

Daughter tubers infected via stolon

Soft rot in store

SUMMER

A 'gappy' crop

WINTER

Bacteria overwinter in lenticels of seed tubers

Bacteria spread to developing shoots

SPRING

daughter tubers in the absence of above-ground symptoms. One of the main reasons that around 70% of European seed tubers are of Dutch origin is that Scottish and other seed producers have more difficulty controlling blackleg in seed crops.

(vi) *Control*

Cultural control is the main approach to reduction of blackleg and soft rot. Certified seed should be bought; this, as described above, will not guarantee a disease-free crop, but should reduce the disease. Crops should not be harvested during wet weather and damage should be kept to a minimum. In crops where there is a low risk of bacterial infection, 'curing' or 'healing' potatoes, by subjecting them to relatively high temperatures (*c.* 16 °C) and 80% relative humidity for 10 days immediately after harvest may help protect potatoes from soft rot. Otherwise cool (*c.* 5 °C), well ventilated and dry potato stores will reduce the considerable damage that can be done. Tubers which have been rained on or are showing any signs of soft rot should not be bulk stored, but marketed as soon as possible. Hot water treatments have given some success in controlling the disease in seed tubers, but they are costly and still at an experimental stage (Wale *et al.*, 1986).

Ring rot

(i) *Causal organism*
Bacterium, *Corynebacterium sepedonicum*.

(ii) *Host range*
The only economically important host is potato.

(iii) *Symptoms*

(a) *on haulm*. Ring rot symptoms on the haulm are manifested as a wilting and rolling of leaves. If infected stems are cut, a milky bacterial ooze can be squeezed out of xylem vessels.

(b) *on tubers*. Tuber symptoms can develop in the field and in store. The vascular ring in affected tubers breaks down and creamy cheese-like bacterial ooze can be squeezed out. It is common for secondary bacterial pathogens to invade tubers colonised by *C. sepedonicum* and tubers are then quickly reduced to a semi-liquid state.

Figure 9.5. Disease cycle of black leg and soft rot (*Erwinia carotovora* subsp. *atroseptica*).

(iv) *Disease cycle*

Diseased seed is the main initial source of inoculum. Seed tubers become infected when they are brought into contact with contaminated machinery, containers and cutting knives. Wounds encourage the disease to develop. In the growing crop, daughter tubers become contaminated via the vascular system and infection is most rapid at soil temperatures around 20 °C.

(v) *Significance*

As with blackleg or soft rot, infection of tubers, whether in the field or in store, is the most significant phase of ring rot. Yield losses in some areas of the USA and Canada have reached 15% of the total crop although recently more effective control measures have reduced the problem somewhat. The disease is of little significance in Europe and has not been reported in the UK, but there are no apparent climatic constraints on its development.

(vi) *Control*

Plant disease legislation has been effective in keeping ring rot out of the UK. As a non-indigenous notifiable disease, growers must report any outbreaks, and consignments of imported potatoes are inspected for ring rot.

Cultural control methods have been effective at reducing ring rot in countries where it is a problem. Certified disease-free seed should be grown and special care taken in cleaning machinery, stores and bulk seed containers. New bags are used for transporting and marketing seed. Good crop rotations should be practised and seed should not be sown in fields where there is a risk of contaminated volunteers emerging.

Brown rot

(i) *Causal organism*
Bacterium, *Pseudomonas solanacearum*.

(ii) *Host range*
All members of the Solanaceae together with many other crops including banana, groundnut and numerous weed species.

(iii) *Symptoms*

(a) *on haulm*. Brown rot symptoms on potato haulms are manifested as wilting, stunting and yellowing of foliage. Transient wilting during the day, with recovery at night, often leads to a permanent wilt and death soon follows. In young potato plants it may be possible to see brown colonised xylem vessels through the epidermis. Cut stems freely ooze greyish-white bacterial slime from xylem vessels.

(b) *on tuber*. The first symptom of brown rot in tubers is a browning of the vascular ring. If squeezed, pale yellow bacterial ooze is visible. This may contaminate the surface of the tuber, especially around the stem and eyes, resulting in soil adhering to the tuber surface.

(iv) *Disease cycle*
The primary initial source of inoculum is mildly or latently infected seed tubers; however, weed species may also harbour the pathogen during winter. The disease is particularly severe in hotter climates and it rarely occurs in areas where mean soil temperatures are below 15 °C.

(v) *Significance*
The disease may limit potato production in tropical or sub-tropical areas, but it rarely causes problems in Europe and has not been reported in the UK.

(vi) *Control*
Plant disease legislation has undoubtedly contributed to the absence of the disease in the UK (see ring rot).

Cultural control methods are the only feasible options in areas where the disease is a problem. The use of certified disease-free seed, good crop rotations and elimination of weeds all help to reduce the disease.

Wart

(i) *Causal organism*
Fungus, *Synchytrium endobioticum*.

(ii) *Host range*
Solanaceae only.

(iii) *Symptoms*

Symptoms of wart usually occur on tubers and occasionally on the stem just above ground level in place of leaves. They never occur on roots. Infection of meristematic regions such as eyes results in the development of irregular cauliflower-like outgrowths. These may enlarge as the season progresses, resulting in badly malformed tubers. At the end of the season, warted tissue darkens and rots.

(iv) *Disease cycle*

The fungus overwinters as resting spores in the soil or on the surface of seed tubers. Upon germination, motile zoospores are released which swim in soil moisture, encyst and penetrate the tuber surface particularly in the region of meristematic tissue. Tuber cells are then stimulated to multiply and warts begin to form. There may be further cycles of infection within the growing season as sporangia are produced in infected tissue which release more infective zoospores. Towards the end of the season zoospores fuse in pairs, resulting in the formation of resting spores which are released into the soil when the warted tissue rots. Such resting spores may remain viable for 30 years. Wet, poorly drained soil together with average early summer temperatures in the UK are ideal for disease development. Dry weather tends to inhibit the disease.

(v) *Significance*

The disease which was once widespread in the UK is now of minor importance as a result of effective control measures. Badly warted potatoes are unmarketable in most parts of the world and, in the UK, wart is a notifiable disease.

(vi) *Control*

Plant disease legislation has played a major role in control of the disease in a number of countries including the UK (Table 9.2).

Disease resistance is the other major measure for control of wart. Most popular varieties of potato in the UK are immune to wart disease. One notable exception to this is the variety King Edward, which should not be grown on wart-infested land.

Powdery scab

(i) *Causal organism*

Fungus, *Spongospora subterranea*.

Table 9.2. *Main requirements of plant disease legislation relating to wart disease of potatoes in the UK*

1. Any occupier of land on which wart disease exists or appears to exist must report the occurrence to the Ministry at once or to an Inspector of the Ministry. This must be done even if the disease on the land has been reported in a previous year. The Minister has the power to declare and define by notice land infested with wart disease and a safety zone surrounding the contaminated plot.
2. The notice served by the Minister will prohibit the planting of any potatoes and the removal of plants for transplanting elsewhere from land infested with wart disease and will restrict potato planting within the safety zone to varieties immune to wart disease.
3. Potatoes visibly affected with wart disease may not be sold or offered for sale for any purpose, and no potatoes from any crop in which wart disease has been found to exist may be sold or offered for sale or planting.
4. The destruction of infected potatoes must be carried out in accordance with the terms of a notice served by an authorised officer of the Ministry.
5. Land on which wart disease has occurred and which is subject to a notice served under the Plant Health (Great Britain) Order 1987 will continue to be affected by the restrictions on planting made under that Order.

(After Brenchley & Wilcox, 1979.)

(ii) *Host range*
Confined to members of Solanaceae.

(iii) *Symptoms (Figure 9.6)*
Symptoms of powdery scab on tubers are initially seen as small, slightly raised pimples under the surface of the skin. The skin then breaks away leaving a ragged edge and a mass of brown powdery spore balls which distinguish these primary symptoms from those of common scab (*Streptomyces scabies*). A more serious phase of powdery scab may develop, especially in wet soils. Tubers become deformed and wart-like growths may develop on tubers and, unlike wart disease, on roots as well.

(iv) *Disease cycle (Figure 9.7)*
The pathogen overwinters as spore balls in soil and on the surface of seed tubers. These germinate to form motile zoospores which swim in soil moisture and invade root hairs, epidermal cells, lenticels or eyes, or penetrate through wounds in the tuber. Plasmodia are formed in tissue and may stimulate the tuber to produce a protective layer of cork which bounds the plasmodia. Secondary zoospores are

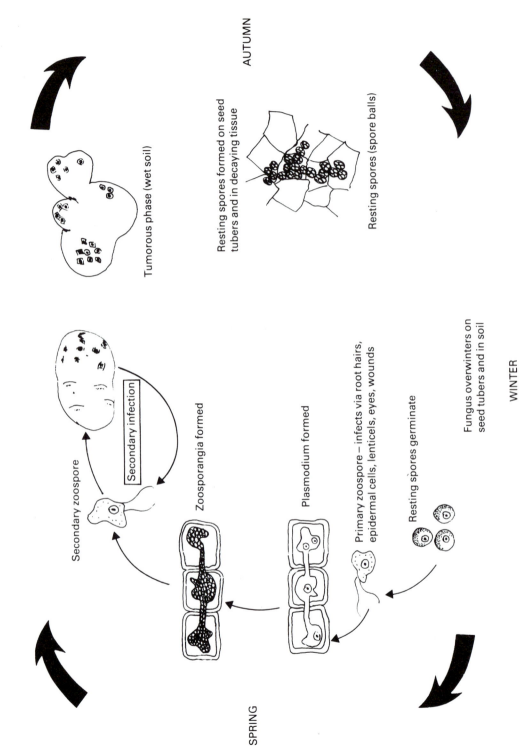

AUTUMN

Resting spores formed on seed tubers and in decaying tissue

Resting spores (spore balls)

WINTER

Fungus overwinters on seed tubers and in soil

Resting spores germinate

Primary zoospore – infects via root hairs, epidermal cells, lenticels, eyes, wounds

Plasmodium formed

Zoosporangia formed

SPRING

Secondary zoospore

Secondary infection

SUMMER

Tumorous phase (wet soil)

Figure 9.6. Powdery scab of potato tubers caused by *Spongospora subterranea*. (Plant Breeding International Cambridge.)

formed from the plasmodia which spread the disease deeper into tissue and can also initiate fresh infections. This is the most destructive phase of the disease. Spore balls are also formed which contaminate soil and can remain viable for 6 years. The disease is worse in wet, poorly drained soils and is frequently observed after two or more periods of 2–3 days continuous rain which occur at 10–14 day intervals in an otherwise cool dry summer.

(v) *Significance*

The disease is seasonal and generally confined to fields with a history of the problem. When it does occur, mild disease attacks reduce marketability of the crop and severe, cankerous powdery scab often results in rejection of the crop. *Spongospora subterranea* is a vector for potato mop top virus (see spraing).

(vi) *Control*

Cultural control of the disease requires the practice of long rotations, improvement in drainage and destruction of infected tubers. Diseased tubers should not be fed to stock as manure will be contaminated with viable spore balls. Certified seed tubers should be grown.

Figure 9.7. Disease cycle of powdery scab (*Spongospora subterranea*).

Disease resistance is available in some of the popular potato varieties. King Edward is one of the most resistant maincrop varieties whereas Pentland Crown carries a much higher risk of infection.

Chemical control. Maneb + zinc oxide (Mazin) applied as a dust at planting is approved for the control of powdery scab.

Common scab

(i) *Causal organism*
Actinomycete, *Streptomyces scabies*.

(ii) *Host range*
Potato, red beet, sugar beet, radish, carrot, turnip.

(iii) *Symptoms (Figure 4.4, p. 72)*
Symptoms of common scab are most frequently seen on mature tubers. Scabs consist of circular or angular lesions on the tuber surface which may be discrete or coalesce to form large patches of scabbed tissue. Scabs are generally quite superficial but in some varieties they may appear raised, or penetrate the tuber to a few millimetres. Common scab gives the potato an overall scruffy appearance.

(iv) *Disease cycle*
The organism is widely distributed throughout soil and penetration of the developing tubers usually occurs via lenticels. Penetration stimulates the tuber to produce a corky barrier to protect itself from further invasion and hence a scab is formed. Sometimes *S. scabies* can penetrate this barrier and a further barrier develops in the tuber giving a deeper scab. Development of scabs rarely occurs in store. The disease is seasonal and tends to be most severe in light, freely drained alkaline soils after periods of dry summer weather. Temperatures of around 20 °C are optimal for disease development.

(v) *Significance*
Common scab is probably the most common disease of potato tubers in the UK and around 4% of the national crop may be affected after a dry June or July. Yield losses have occasionally been reported, but the disease is most significant because it reduces quality of tubers. Most scabs can be removed by peeling potatoes but the cosmetic appearance is affected, especially when sold in

transparent packages. Supermarket buyers generally reject tubers with any quantity of scab. Scab on seed crops can result in rejection of the crop as seed.

(vi) *Control*
Cultural control of scab involves using certified disease-free seed, and irrigating during early stages of tuber formation. Green manuring, i.e. incorporating large amounts of organic material such as grass, rape or soybeans may be beneficial if the population of *S. scabies* is not too great in soil.

Disease resistance is available in currently-grown varieties (NIAB Farmers' Leaflets No. 3, *Recommended Varieties of Potatoes*). Maris Piper and Desirée are among the more susceptible varieties whereas Pentland Crown is much more resistant. Susceptible varieties should not be grown on soils conducive to the disease.

Pink rot

(i) *Causal organism*
Fungus, *Phytophthora erythroseptica*.

(ii) *Host range*
Potato, tomato, clover, asparagus, raspberry.

(iii) *Symptoms*
The fungus attacks tubers and roots and it is possible for diseased plants to wilt in warm weather. Affected tubers tend to have dark lenticels and a rubbery texture. If squeezed, they exude a watery fluid and it is common for particles of soil to stick to diseased tubers. When tubers are cut open affected tissue is initially an off-white colour. However, within 30 minutes, the tissue turns a salmon-pink colour and eventually purple-brown or black. Diseased tubers often smell of vinegar.

(iv) *Disease cycle*
The pathogen survives for many years in infested soil as oospores. Upon germination, mycelium or zoospores infect all below-ground parts of the potato plant. Rotten infected daughter tubers release more oospores into soil. The disease cannot spread to healthy undamaged tubers in store, but in poor stores, wounded potatoes may be susceptible to infection. Pink rot tends to be most severe in heavy, badly drained soils.

(v) *Significance*

The disease tends to be seasonal and restricted to fields with a history of the problem. However, secondary bacterial colonisation of affected tubers may mask the primary pathogen and therefore pink rot attacks may be underestimated.

(vi) *Control*

Cultural control includes crop rotation and improvement in drainage.

The remainder of the potato diseases described, except virus diseases, occur primarily on the tuber after harvest.

Watery wound rot (leak)

(i) *Causal organism*
Fungus, *Pythium ultimum* (*Pythium debaryanum*).

(ii) *Host range*
The fungus causes damping-off and root rot in a wide range of crops as well as rot of potato tubers.

(iii) *Symptoms*
Watery wound rot appears to be unable to infect undamaged tubers either in the field or in store. However, if tubers are subjected to mechanical damage or bruising during harvesting, the fungus produces a discoloured water-soaked area around the damaged site. If tubers are cut open, a dark brown line separates diseased and healthy tissue. Affected parts of the tuber deteriorate quickly and in most cases secondary bacterial invasion leads to total collapse of the tuber. It is difficult in the latter phase of the disease to determine the primary cause as both soft rot and, to a lesser extent pink rot, may cause similar symptoms.

(iv) *Disease cycle*
Fungal oospores in soil, in the field, or adhering to harvested tubers germinate and penetrate tubers through damaged tissue. More infection sites may be initiated in store in infected tubers by the production of sporangia. Rot may then progress quickly, especially at temperatures around 21 °C. Tubers left to rot in the field will further contaminate soil with oospores.

(v) *Significance*
The disease is only occasionally serious but it may be the initiator of more serious problems such as soft rot, which occur in potato stores.

(vi) *Control*
Cultural control is the main way by which the disease can be controlled. Tubers left in the soil should, as far as possible, be collected and damage at harvest reduced to a minimum. Rotten potatoes should not be put into store, or returned to land.

Gangrene

(i) *Causal organism*
Fungus, *Phoma exigua* var. *foveata*.

(ii) *Host range*
Potatoes.

(iii) *Symptoms (Figure 9.8)*
Gangrene is a storage disease of potatoes which is not normally seen until potatoes have been stored for at least a month. First symptoms of the disease are small dark round or oval depressions in the tuber surface. These may be associated with wounds, eyes or lenticels. Lesions gradually enlarge giving characteristic 'thumb-mark' depressions covered by smooth darkened skin. It may be possible to see small dark pycnidia in lesions. When affected tubers are cut open, large cavities lined with white fungal mycelium may be seen inside. There is a clear distinction between healthy and colonised tissue (cf. dry rot).

Figure 9.8. Gangrene of potato tubers caused by *Phoma exigua* var. *foveata*. (P. Gans, NIAB.)

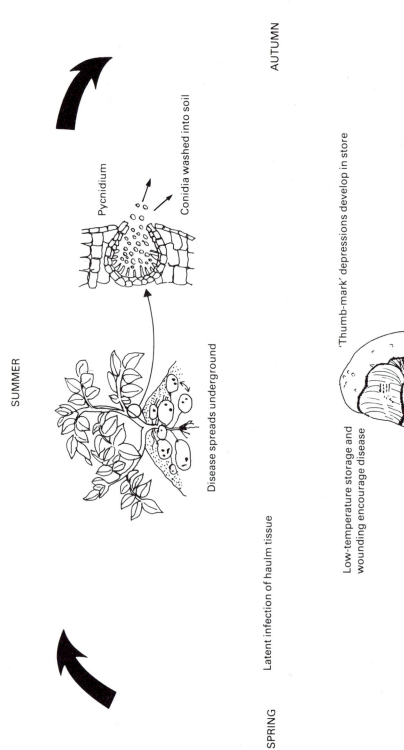

AUTUMN

Pycnidium

Conidia washed into soil

Disease spreads underground

'Thumb-mark' depressions develop in store

SUMMER

WINTER

Low-temperature storage and wounding encourage disease

Fungus overwinters in soil and on seed tubers

Latent infection of haulm tissue

SPRING

(iv) *Disease cycle (Figure 9.9)*

The fungus overwinters in soil either in the field or in potato stores. Seed tubers and groundkeepers also serve as overwintering sources of inoculum. Infection of potato haulm tissue and developing tubers occurs during the season, but goes largely unnoticed. Pycnidia may be formed on senescent tissue at the end of the season and spores released can be washed through the soil and contaminate tubers. Undamaged tubers can be infected by the fungus, but infections may be latent and require tuber damage before progressing. High temperatures in late summer tend to reduce the disease and low temperature storage ($2-6\,°C$), which inhibits the healing process in potatoes, encourages disease development.

(v) *Significance*

A very high degree of contamination is required to reduce the yield of potatoes in the field significantly. The disease causes more problems in stored ware and seed potatoes.

(vi) *Control*

Cultural control methods include early lifting, reduction of damage during harvest and healing potato crops by subjecting them to a temperature of $16\,°C$ and 80% relative humidity for 10 days immediately after harvest. Potatoes should then be stored at *c.* $5-8\,°C$. Groundkeepers and debris should be eliminated from fields.

Disease resistance is available in currently grown varieties, although no particular variety is outstanding in this respect. Varieties differ in their susceptibility to damage during harvesting with consequent differences in susceptibility to most storage diseases, including gangrene.

Chemical control is desirable in crops intended for seed. Fungicides should be applied at store loading or as soon as possible afterwards. Fungicides approved for the control of gangrene include:

Active ingredient	*Product example*
imazalil	Fungaflor C
iodophor + thiabendazole	Byatran, Tubazole
tecnazene + thiabendazole	Hytec Super, Storite SS
thiabendazole	Storite, Tecto
2-aminobutane	CSC 2-aminobutane (by specialist contractor only)

Figure 9.9. Disease cycle of gangrene (*Phoma exigua* var. *foveata*).

Dry rot

(i) *Causal organism*
Fungi, various *Fusarium* spp. including *F. coeruleum* (*F. solani* var. *coeruleum*) and *F. sulphureum*.

(ii) *Host range*
Both species are important mainly as pathogens of potatoes.

(iii) *Symptoms*
Symptoms of dry rot usually occur in tubers which have been stored for a number of weeks. Initially, small brown areas may be visible on the tuber surface. These patches may be associated with wounds or damage caused by other pathogens such as *P. infestans*. Lesions enlarge and the tuber skin becomes wrinkled. White, orange or blue fungal colonies then develop on the tuber surface. If diseased tubers are cut open, internal tissue appears brown and large cavities lined with white or grey fungal mycelium may develop. There is no clear distinction between colonised and healthy tissue. Tubers colonised by the dry rot pathogens are susceptible to invasion by bacterial soft rots.

(iv) *Disease cycle*
Fusarium species survive in the soil as resting spores (chlamydospores). Seed tubers may also be contaminated with chlamydospores. Infection usually occurs as a result of damage to tubers and progresses well in poorly ventilated humid stores. There is some debate as to the optimum temperature requirement for the disease; *Fusarium* spp. will grow over a wide temperature range. Should partially rotted seed tubers be re-planted, the rot will progress rapidly in the field and more chlamydospore inoculum will be released into the soil.

(v) *Significance*
In the early days of the mechanisation of potato harvesting, there was a dramatic increase in dry rot in stores because of the damage done by unrefined potato harvesters; this is much less of a problem now. However, some seed crops still have problems with dry rot, mainly because of the extra handling and storage involved in their production. Some older varieties are very susceptible to the disease.

(vi) *Control*

Cultural control of dry rot involves minimising damage at harvesting and during subsequent handling. Well ventilated cool stores will also help to reduce the disease.

Chemical control, particularly for seed tubers, may be worthwhile. Products should be applied as soon as possible after lifting. Fungicides approved for control of dry rot include:

Active ingredient	Product example
carbendazim + tecnazene	Hortag Tecnacarb Dust
iodophor + tecnazene	Bygran F
iodophor + thiabendazole	Byatran, Tubazole
imazalil	Fungaflor C
tecnazene	Fusarex, Hytec
tecnazene + thiabendazole	Hytec Super, Storite SS
thiabendazole	Storite, Tecto

Note: special precautions should be taken when using products containing tecnazene, a sprout suppressant, on seed tubers. Consult relevant agrochemical product manual.

Skin spot

(i) *Causal organism*
Fungus, *Polyscytalum pustulans*.

(ii) *Host range*
Solanaceae, most significantly potatoes.

(iii) *Symptoms*
The fungus can develop on all underground parts of the potato plant, giving rise to general browning. However, infections of the tuber during storage are most significant. After several months in store, small discrete barnacle-like pimples occur on the tuber surface. They may be dark and usually have a dark ring at the base. Skin spot may develop over the entire tuber surface, including the eyes.

(iv) *Disease cycle*
The fungus can overwinter in soil as microsclerotia and in dry soil in potato stores. Diseased seed tubers, however, are the main initial source of inoculum. Infection spreads to underground parts of the plant throughout the season and is usually concentrated around the eyes. Affected tubers are generally symptomless at harvest and only

after storage do spots develop. Wet soils during lifting and low store temperatures immediately after lifting appear to favour the disease. Damp conditions in potato stores can result in further infections of tubers via air-borne conidia.

(v) *Significance*

The disease may be important in two respects. Firstly, the cosmetic quality of a crop is reduced as a result of skin spot infection, and secondly, colonisation of potato eyes can reduce sprout numbers resulting in non-emergence of plants.

(vi) *Control*

Cultural control methods are important. Potatoes should be lifted early during warm conditions and stored in dry, well ventilated stores. Seed potatoes should be sprouted prior to planting and those which fail to sprout should be rejected.

Disease resistance is available in a number of commercially grown varieties in the UK, including Desirée and Pentland Squire.

Chemical control may be worthwhile in tubers for seed. Fungicides should be applied as soon as possible after lifting. Products approved for the control of skin spot include:

Active ingredient	Product example
imazalil	Fungaflor C
iodophor + tecnazene	Bygran F
iodophor + thiabendazole	Byatran, Tubazole
tecnazene + thiabendazole	Hytec super, Storite SS
thiabendazole	Storite, Tecto
2-aminobutane	CSC 2-aminobutane (applied by specialist contractor)

Silver scurf

(i) *Causal organism*
Fungus, *Helminthosporium solani*.

(ii) *Host range*
Potatoes only.

(iii) *Symptoms (Figure 9.10)*

Symptoms of silver scurf develop on tubers mainly during storage. Superficial grey or silvery patches occur on the tuber surface almost as if the skin has been abraded slightly. During storage, particularly in humid conditions, lesions may cover the entire tuber surface and take on a sooty appearance. Increased permeability of affected skin can result in potatoes drying out and shrivelling.

(iv) *Disease cycle*

The fungus overwinters on tubers left in soil and on seed tubers. Penetration of the tuber may occur via lenticels, or directly through the periderm, but the fungus is confined to the outer layers of the tuber. Spores which are produced on affected tubers, resulting in the sooty appearance, may be dispersed to healthy tubers. Warm (15–20 °C), humid potato stores encourage disease development.

(v) *Significance*

The disease is very widespread but because of its superficial nature is relatively unimportant in most crops. However, it can reduce the marketability of a crop because of its detrimental effect on the appearance of diseased tubers and loss of water in storage.

(vi) *Control*

Cultural control methods to reduce silver scurf include early harvesting and cool dry storage conditions.

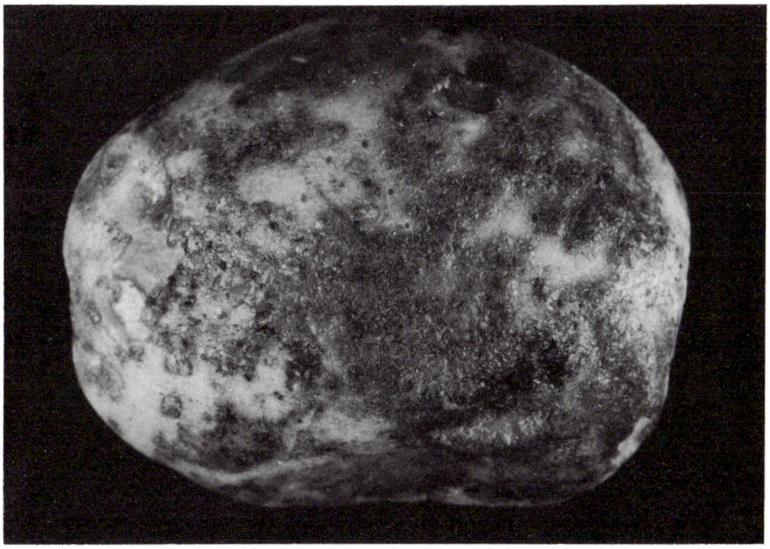

Figure 9.10. Silver scurf of potato tubers caused by *Helminthosporium solani*. (Purdue University.)

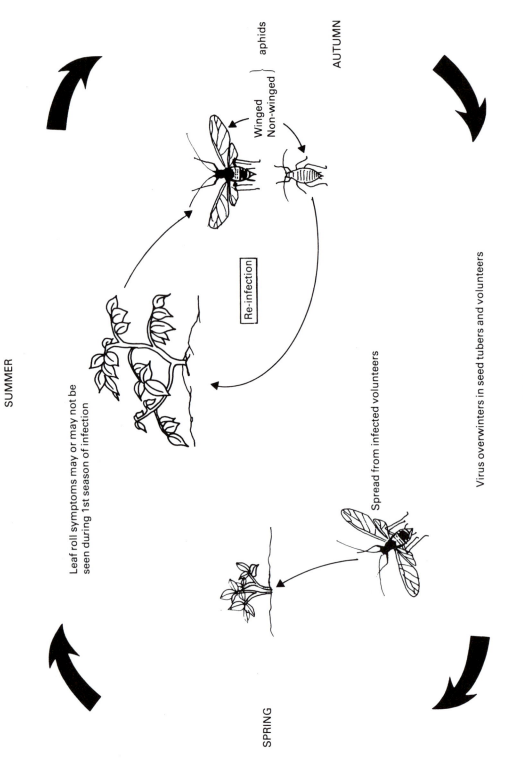

AUTUMN

Winged
Non-winged } aphids

Re-infection

WINTER

Virus overwinters in seed tubers and volunteers

SUMMER

Leaf roll symptoms may or may not be seen during 1st season of infection

Spread from infected volunteers

SPRING

Chemical control is worthwhile only when growing for top quality markets, long storage and in some seed crops. Fungicides should be applied as soon as possible after lifting. Products approved for the control of silver scurf are the same as those approved for control of gangrene.

Leaf roll

(i) *Causal organism*
Virus, potato leaf roll virus (PLRV). Vectors are aphid species, most notably *Myzus persicae*.

(ii) *Host range*
Mainly Solanaceae.

(iii) *Symptoms*
There are two disease syndromes. Should infection occur early in the season, by mid-July the top leaves on affected plants turn yellow, appear erect and their margins roll inwards. The virus can spread to the tubers during the first season of the disease whether or not these first foliar symptoms are visible. The second and more severe syndrome occurs in plants grown from contaminated tubers. Generally after about one month's growth diseased plants appear erect and stunted. Top and basal leaves are faintly chlorotic and rolled. Basal leaves may also become stiff and leathery.

(iv) *Disease cycle (Figure 9.11)*
Primary symptoms occur after plants have been infected by viruliferous aphids (*Myzus persicae*) spreading from diseased volunteer plants and plants arising from contaminated tubers. Virus transmission is of the persistent (circulative) type and the aphid needs to feed on a diseased plant for about two hours before it acquires the virus; a 24-hour incubation period is needed before transmission. Contaminated aphids remain so for many days.

(v) *Significance*
PLRV is the most common virus disease of potatoes and because of its severity it is also the most important. In severely diseased crops, yield losses of up to 80% have been reported. Annual average world-wide losses may be as much as 10%.

Figure 9.11. Disease cycle of leaf roll (potato leaf roll virus). Vector: *Myzus persicae*.

(vi) *Control*

Cultural control rests mainly on the use of certified seed potatoes and the elimination of volunteers. Seed crops are grown in areas where aphids are less common and should be harvested early. It is possible, although uneconomic on a large scale, to free contaminated tubers by heating (25 days at 37.5 °C). The highest grade of seed is maintained from stem cuttings made from micropropagated stocks which are checked in the early years of seed multiplication by virus testing (virus tested stem cuttings, VTSC).

Disease resistance is available in currently grown potato varieties. Pentland Crown is one of the more resistant whereas Romano is susceptible.

Chemical control is possible for the aphid vector of PLRV and may provide worthwhile reduction of virus spread in seed potato crops. Products approved for the control of aphids in potatoes include:

Active ingredient	*Product example*
aldicarb	Temik 10G
deltamethrin + heptenophos	Decisquick
demeton-S-methyl	Metasystox 55
dimethoate	Dimethoate
disulfoton	Disulfoton, Disyston P-10
malathion	Malathion 60
malathion + pyrethrins	Cromocide
oxamyl	Vydate 10G
oxydemeton − methyl	Metasystox R
phorate	Phorate
pirimicarb	Aphox
thiofanox	Dacamox 5G
thiometon	Ekatin

Strains of *Myzus persicae* have developed resistance to all commercially available aphicides.

Many of the above compounds must be soil-incorporated.

Mild mosaic

(i) *Causal organism*

Virus, mainly potato virus X (PVX). The virus is mechanically transmitted.

(ii) *Host range*

The main natural hosts of PVX are members of the Solanaceae.

(iii) *Symptoms*

Symptoms of PVX may be latent. A mild mosaic may be visible, particularly if there is cloudy weather and a white card is held behind the leaf. Necrosis has been attributed to infection by PVX in artificial inoculation tests.

(iv) *Disease cycle*

The main initial source of inoculum is seed tubers. Infections in haulm tissue are transmitted by machinery, animals and plant-to-plant contact.

(v) *Significance*

Until the use of VTSC, PVX was the most common virus disease on potatoes. Very occasionally yield losses of up to 50% were recorded as a result of infection by PVX. However, PVX on its own is today of little economic importance in potato crops. It is of more significance as a component of double virus infections, e.g. PVX + PVY results in a severe mosaic symptom.

(vi) *Control*

The main control method is to use certified seed.

Severe mosaic, rugose mosaic, leaf drop streak

(i) *Causal organism*

Virus, potato virus Y (PVY). Vectors are aphid species, most notably *Myzus persicae*.

(ii) *Host range*

Solanaceae.

(iii) *Symptoms*

As with PLRV, PVY causes progressive symptoms over two seasons. Primary (first year) infections result in very variable symptoms, commonly dark spots or streaks on leaf veins, or a mild mosaic or mottle. Sometimes leaves shrivel and drop (leaf drop streak), but generally symptoms are slight. Planting diseased tubers results in more severe symptoms during the following season. Plants are often stunted and have brittle wrinkled leaves. Severe mosaic, rugose mosaic and leaf drop streak symptoms may also occur. There is generally no tuber symptom.

(iv) *Disease cycle*

The disease cycle is very similar to PLRV except that virus transmission is non-persistent. The main vector is *Myzus persicae*.

(v) *Significance*

Yield losses as a result of PVY may reach 80%. The disease is probably the next most significant virus disease after PLRV.

(vi) *Control*

Cultural control of PVY involves planting certified seed, isolation of seed potato crops and early lifting.

Disease resistance is available in currently grown varieties, for example Pentland Crown; King Edward is susceptible.

Chemical control of the vector is probably not worthwhile because of the non-persistent mode of transmission.

Spraing (TRV)

(i) *Causal organism*

Two viruses cause spraing, tobacco rattle virus (TRV) which is the most important, and potato mop top virus (PMTV). The vectors of TRV are free-living, soil-borne nematodes of the genera *Trichodorus* and *Paratrichodorus*.

(ii) *Host range*

More than 100 plant species may be infected by TRV.

(iii) *Symptoms*

No symptoms develop on the haulm in the season during which the tubers become infected from the soil. However, plants grown from TRV-affected tubers the following season show a range of symptoms from unusual chlorotic or corky lines on leaves to no symptoms at all. In the tuber, TRV infection gives rise to brown corky arcs which are referred to as spraing (Scottish dialect word for streaks or stripes). These are generally not visible from the tuber surface.

(iv) *Disease cycle*

Nematode vectors become contaminated with TRV by feeding on the many potential infected hosts. They can remain infective for as long as 5 years, but the virus is not transmitted to offspring. Tubers may become infected directly if viruliferous nematodes feed on them. Spraing (TRV) is particularly common on light sandy or gravelly soils during rainy seasons.

(v) *Significance*
Affected tubers are unmarketable, although the disease is not particularly widespread in the UK.

(vi) *Control*
Disease resistance is available in currently grown varieties. Romano is quite resistant, whereas Pentland Dell is very susceptible.

Chemical control of the nematode vectors of TRV is possible by incorporation of pesticides such as aldicarb (Temik 10G) or oxamyl (Vydate 10G) into soil.

Spraing (PMTV)

(i) *Causal organism*
Virus, potato mop top virus (PMTV). The vector is the soil-borne fungus *Spongospora subterranea* which causes powdery scab of potatoes.

(ii) *Host range*
Members of Solanaceae.

(iii) *Symptoms*
Like TRV, infection by PMTV does not usually result in symptoms on the haulm in the season during which the tubers become infected. Plants grown from PMTV-affected tubers the following season may also be symptomless, but can develop symptoms very similar to those produced by TRV. Alternatively, in some varieties 'mop tops' develop, i.e. stems with short internodes and crowded leaves. Infection of tubers results in spraing. The severity of spraing varies according to variety. Generally spraing caused by PMTV can be seen more clearly from the tuber surface.

(iv) *Disease cycle*
See disease cycle of vector, *S. subterranea* (potato powdery scab). Unlike spraing caused by TRV, PMTV is not found predominantly on any one soil type.

(v) *Significance*
As TRV.

(vi) *Control*
Disease resistance is available in currently grown varieties. Some varieties, e.g. Maris Piper and King Edward rarely develop spraing whereas Pentland Crown is much more likely to develop the disease. Control methods used to control *S. subterranea* may also be useful in reducing spraing.

Further reading
Anon. (1985). *Potato Diseases*. Potato Marketing Board (Oxford)/
    NIAB (Cambridge) publication.
Brenchley, G. H. & Wilcox, H. J. (1979). *Potato Diseases*. ADAS
    RPDl MAFF publication. London: HMSO.
Hooker, W. J. (ed.) (1981). *Compendium of Potato Diseases*. St Paul:
    The American Phytopathological Society.
Ministry of Agriculture, Fisheries and Food (1984). *Control of Diseases
    of Potatoes*. ADAS Booklet 2388. Alnwick, Northumberland: MAFF
    Publications.
National Institute of Agricultural Botany (1989). Farmers' Leaflet
    No. 3, *Recommended Varieties of Potatoes*. Cambridge: NIAB
    publication.

10 Diseases of sugar beet

Sugar beet is grown in most European countries and North America. The 200 000 ha of sugar beet grown in the UK is concentrated around processing factories. Sugar beet production is organised on a quota system by the processor British Sugar and the area grown has remained static for a number of years. Approximately one half of the UK sugar requirement is met by UK grown beet.

The most significant diseases of sugar beet are caused by viruses. Rhizomania disease occurs in many sugar beet growing areas of the world and poses the greatest potential threat to beet production in the UK should it become established.

Blackleg (damping-off)

(i) *Causal organism*
Fungi, including *Pythium* spp., *Aphanomyces cochlioides* and *Rhizoctonia solani*. *Phoma betae* may also cause pre-emergence damping-off (see *Phoma* diseases).

(ii) *Host range*
Pythium spp. and *Rhizoctonia solani* cause damping-off in a wide range of crops. *Aphanomyces cochlioides* is relatively specific to beet.

(iii) *Symptoms*
The first symptoms of blackleg on beet seedlings are irregular brown spots on the hypocotyl and root. These lesions enlarge, resulting in thin brown shrivelled roots. Severely diseased seedlings usually

collapse and die. Less severely affected plants may continue to grow but are stunted and may produce poor yields of deformed beet.

(iv) *Disease cycle*
This varies according to the fungal species. *Pythium* spp., *A. cochlioides* and *R. solani* are primarily soil-borne pathogens. Blackleg tends to be more of a problem in poorly drained acid soils.

(v) *Significance*
Although in the past, losses due to blackleg were sufficient to cause crop failure, currently the disease is well controlled by fungicide seed treatments. Blackleg caused by *A. cochlioides* can be severe in crops sown late into soils with high inoculum levels.

(vi) *Control*
Cultural control. Well timed sowings into well prepared seedbeds of adequate soil nutrition will reduce the problem. Application of lime together with improved drainage will also help.

Chemical control. All sugar beet seed supplied and grown under contract to British Sugar is fungicide-treated. Tachigaren (hymexazol) is incorporated into the seed pellet to control soil-borne damping-off fungi. Thiram (Tripomol 80) is also approved for control of damping-off.

Phoma diseases

(i) *Causal organism*
Fungus, *Phoma betae* (*Pleospora bjoerlingii*).

(ii) *Host range*
Mainly beet.

(iii) *Symptoms*
The pathogen can attack beet at all stages of its development. Initially *P. betae* causes blackleg (pre-emergence damping-off). Mature leaves of beet can also be infected by *Phoma*, causing leaf spot. The resulting light brown circular lesions contain black pycnidia arranged in concentric rings near the lesion edge. *Phoma betae* is also one of the major causes of storage rots of beet in the USA. After storage for about 80 days or more the rot usually begins in the centre of the crown and spreads downwards into the beet,

resulting in a cone-shaped area of black rotten tissue. Pockets lined with white mycelium may develop and black pycnidia are also formed.

(iv) *Disease cycle*
The fungus overwinters on seed and on debris. The pycnidia formed release spores during wet weather, which are spread mainly by rain-splash and insects. Spread of the pathogen in seed crops may also be systemic.

(v) *Significance*
In the UK, *P. betae* is probably the most important of the blackleg pathogens. The leaf spot phase of the disease is of little economic importance except as a source of inoculum for seeds and roots. *Phoma betae* is important as a storage rot of roots, particularly in the USA where periods of storage may be as long as 150 days.

(vi) *Control (see blackleg)*
Chemical control of blackleg caused by *P. betae* also involves steeping the seed in thiram (previously diethyl mercuric phosphate, EMP). Crop rotation, mandatory in the UK, will reduce foliar infection. Storage rots are reduced in the USA by application of thiabendazole, avoiding wounds at harvesting and allowing beet to 'heal' before being put into piles.

Alternaria leaf spot

(i) *Causal organism*
Fungus, *Alternaria alternata* (*Alternaria tenuis*).

(ii) *Host range*
Numerous.

(iii) *Symptoms*
Symptoms appear late in the growing season. Leaves turn brown from the edge inwards. Severe attacks can kill leaves. Characteristic *Alternaria* target spots may be seen. *Alternaria* may rot beet roots stored in clamps.

(iv) *Disease cycle*
The fungus is weakly pathogenic and can attack only senescing tissue. Its large club-shaped spores are wind-blown onto beet from any contaminated decaying plant material.

(v) *Significance*
Alternaria leaf spot is quite widespread in beet crops prior to harvest but is generally unimportant. It is commonly associated with virus yellows infected plants where it may contribute to the yield loss associated with this disease. It may also contribute to the general end-of-season losses attributable to a combination of pathogens.

(vi) *Control*
Control of the disease is currently not considered worthwhile.

Downy mildew

(i) *Causal organism*
Fungus, *Peronospora farinosa* f.sp. *betae*.

(ii) *Host range*
Confined to beet.

(iii) *Symptoms*
Symptoms usually occur on the central heart leaves of sugar beet plants. Leaves turn light green, distort and thicken. A purple-brown downy sporulating fungal growth then appears. In wet weather the growth may occur on both sides of the leaf, but is generally confined to the lower surface. Older leaves may show signs of severe chlorosis and eventually rot away. Although few plants die as a result of the disease, occasionally the central growing point of the plant is killed. Re-growth often takes place from side shoots.

(iv) *Disease cycle*
The fungus can overwinter as oospores in soil, but the main overwintering source of inoculum is resting mycelium in living green tissue (groundkeepers or seed crops). Sporangia are produced and germinate best at temperatures between 9 and 12 °C during periods of high humidity. The disease is seen most commonly in spring or autumn (seed crops). Mature plants may be more resistant to the disease.

(v) *Significance*
Disease epidemics are sporadic but yield losses of 50% have been reported in badly affected crops. Early attacks tend to reduce root yield whereas late attacks affect sugar content. The disease has declined in recent years in the UK, probably as a result of the change in cultivation of beet crops. Seed-beet crops, the main

overwintering source of inoculum, have tended to become more concentrated in specific areas of the country, away from root crops. There are specified minimum distances between root and seed crops in contracts offered to growers by British Sugar.

(vi) *Control*

Cultural control involves removal and destruction of affected plants and groundkeepers. Crop rotation, mandatory in the UK, will help if soil-borne oospore inoculum is a problem.

Disease resistance is available in currently grown varieties (NIAB Farmers' Leaflet No. 5, *Recommended Varieties of Sugar Beet*).

Powdery mildew

(i) *Causal organism*
Fungus, *Erysiphe betae*.

(ii) *Host range*
Beet species only.

(iii) *Symptoms*
A powdery white growth develops on both leaf surfaces, especially on the outer leaves. At first all that is apparent is that the shine has been taken off affected leaves, but eventually chlorosis occurs and if the attack is severe, leaves may die. Black cleistothecia usually develop on older leaves as the epidemic proceeds.

(iv) *Disease cycle*
The fungus probably overwinters as resting mycelium on groundkeepers, weed beet and seed crops. The role of cleistothecia as overwintering structures in the UK has yet to be established. Conidia produced from resting mycelium initiate infections during late July in the UK. Infection is favoured by temperatures around 20 °C and humid weather. Dry weather then favours colonisation of the plant. Cooler rainy weather slows down a disease epidemic.

(v) *Significance*
Powdery mildew is probably the next most significant disease of sugarbeet, after virus yellows. Early summer attacks can reduce sugar yields by up to 20%. However, the amount of disease has declined in the UK in recent years. This is probably a result of increased use of fungicides to control the disease and a succession of

cold winters which have killed overwintering inoculum (Asher, 1986).

Control
Cultural control primarily involves elimination of groundkeepers and isolation of seed crops from root crops.

Disease resistance is available but generally low in currently grown sugar beet varieties in the UK.

Chemical control of powdery mildew is quite common in the more southerly beet-growing regions of the UK. Fungicides should be applied as soon as symptoms are seen and may be repeated after about 14 days. A single well timed spray is usually sufficient. Sprays after mid-September are seldom cost-effective. Products approved for control of powdery mildew include:

Active ingredient	*Product example*
copper sulphate + sulphur	Top-Cop
sulphur	Kumulus S, Solfa, Thiovit
triadimefon	Bayleton
triadimenol	Bayfidan

Cercospora **leaf spot**

(i) *Causal organism*
Fungus, *Cercospora beticola*.

(ii) *Host range*
Beet species only.

(iii) *Symptoms*
The first symptoms of the disease are small reddish dots on leaves. These develop into small (3–5 mm) circular grey-brown lesions with a red-brown margin. Under conditions of high humidity, groups of grey-black conidiophores develop in the centre of lesions. In severe attacks, lesions spread rapidly over the surface of the leaf and leaves may die.

(iv) *Disease cycle*
The main source of overwintering inoculum occurs on debris, where sclerotia or stromata (a mass of vegetative hyphae) can be formed. Seed crops and seed itself may also carry the pathogen. Conidia

germinate best at high relative humidities (over 90%) and at a relatively high temperature (27 °C). They are dispersed mainly by rain-splash and to some extent, wind.

(v) *Significance*
The fungus is most troublesome in warmer areas of the world where rainfall is high or irrigation is used, e.g. Southern and Central Europe, mid-western USA. It is rarely encountered in the UK. In France yield losses in excess of 40% have been reported in experiments.

(vi) *Control*
Cultural control of the disease involves wide spacing of crops, isolation of seed and root crops and crop rotation. Debris and groundkeepers must also be eliminated. Irrigation should be used only when necessary.

Disease resistance is available in popular varieties grown in countries where there is a problem with the disease.

Chemical control methods are available. Thiabendazole, mixed with one of the dithiocarbamates, appears to be one of the most effective fungicides. There is a problem with MBC-resistant strains of *C. beticola*. Maneb (Ashlade Maneb Flowable) and maneb + zinc (Manex) are approved for control of the disease in the UK.

Forecasting disease is practised to some extent in France. The method is based on the following: (i) observations of primary infection; (ii) temperatures above 17 °C; (iii) 2–3 mm rainfall. Should these three factors occur simultaneously, a fungicide should be applied within 3–5 days.

Ramularia leaf spot

(i) *Causal organism*
Fungus, *Ramularia beticola* (*Ramularia betae*).

(ii) *Host range*
Beet species only.

(iii) *Symptoms (Figure 10.1)*
The disease is generally seen on older leaves towards the end of the season. Initially, small pale spots with a dark border develop. The

lesions enlarge and may develop small white spore masses. *Ramularia* can be distinguished from *Cercospora* by the presence of dark spore masses in lesions of the latter. Lesions often coalesce leading to large areas of necrotic tissue and, in severe attacks, dead leaves.

(iv) *Disease cycle*

The fungus overwinters mainly on debris, although it can be seed-borne. Seed crops and groundkeepers can also act as overwintering hosts. High relative humidity (over 70%) and temperatures around 17 °C are optimal for disease development. The disease is most common from September onwards in the UK.

(v) *Significance*

The disease is widespread in many temperate sugar beet growing areas of the world. However, as a late-season disease in the UK it is usually not highly significant, except in seed crops. Losses between 5 and 10% have been reported in root crops but seed crop losses may be up to 20%.

(vi) *Control*

Cultural control, as for *Cercospora beticola*.

Figure 10.1. *Ramularia* leaf spot of sugar beet caused by *Ramularia beticola*. (M. J. C. Asher, Broom's Barn Experimental Station.)

Chemical control may be worthwhile in seed crops, but no fungicides are currently approved for control of *Ramularia* in the root crop. Fentin hydroxide (Du-Ter 50) is approved for control of the disease in seed crops only but is currently under review with regard to its possible mammalian toxicity. A mix of MBC + a dithiocarbamate may also be effective.

Rust

(i) *Causal organism*
Fungus, *Uromyces betae*.

(ii) *Host range*
Beet species.

(iii) *Symptoms*
The disease is perhaps most obvious during late summer on sugar beet leaves, although early summer infections may also occur. Primary symptoms consist of small red-orange pustules surrounded by chlorotic haloes scattered over both leaf surfaces, sometimes occurring in clusters. Late summer symptoms consist of darker brown, more evenly spread pustules. Severe attacks can give leaves an overall brown appearance and may result in death.

(iv) *Disease cycle*
The fungus overwinters primarily on seed crops or groundkeepers as the teliospore stage (end-of-season symptoms). Teliospores may also contaminate seeds. The optimum temperature for rust development occurs between 15 and 22 °C, and temperatures above 22 °C inhibit rust development.

(v) *Significance*
The disease usually occurs too late in the season to reduce yields of root crops severely. It is more of a problem in seed crops.

(vi) *Control*
Specific control measures are usually not worthwhile although good farm hygiene, e.g. elimination of groundkeepers, will obviously help to reduce the disease. Triadimenol (Spinnaker) is an approved fungicide for control of the disease.

Violet root rot

(i) *Causal organism*
Fungus, *Helicobasidium purpureum* (*Helicobasidium brebissonii*).

(ii) *Host range*
Very wide host range including sugar beet, carrot, potato, lucerne and many agricultural weed species.

(iii) *Symptoms*
In the field the symptoms are circular patches of wilted plants which occur particularly during dry conditions. Examination of affected plant roots reveals violet spots or a network of violet strands which spread to give large purple-brown rotting areas of tissue.

(iv) *Disease cycle*
The fungus overwinters as soil-borne sclerotia or on susceptible weeds. Disease spread in the field may occur by contaminated machinery, wind-dispersal of sclerotia or contaminated manure.

(v) *Significance*
Severe attacks in the field can decrease root yield and sugar content. However, the fungus is probably more significant as the initiator of rots which occur in stored piles (clamps) of beet.

(vi) *Control*
Cultural control methods such as good crop rotations avoiding susceptible crops, elimination of groundkeepers and susceptible weeds and avoiding the spread of disease via contaminated manure will all reduce the disease. Spread in clamps can be avoided by transporting crops to the factory as soon as possible after harvest.

Virus yellows

(i) *Causal organism*
Viruses, beet yellows virus (BYV), beet mild yellowing virus (BMYV) and beet western yellows virus (BWYV). Several different aphid species act as vectors for these viruses, but the most common is *Myzus persicae*.

(ii) *Host range*
The viruses have a wide host range. Cultivated hosts include beet and spinach. BWYV and BMYV can also infect lettuce. Important weed hosts include shepherd's purse, groundsel and chickweed.

(iii) *Symptoms*

Initial foci of the disease appear as small patches of chlorotic plants in fields during summer in the UK. Closer examination of affected plants reveals yellowing of leaves between the veins. Leaves may also become brittle and thicken. Infection by BYV results in the development of small reddish or brown necrotic spots on older leaves which, in combination with the chlorosis, gives plants a bronze cast. Infection by BMYV and BWYV results in an overall golden yellow discoloration developing in the crop. Severe disease epidemics may result in almost all plants showing symptoms. Facultative parasites such as *Alternaria alternata* may grow on affected leaves.

(iv) *Disease cycle*

The viruses overwinter in a variety of susceptible perennial hosts including weed species. Seed crops and beet or mangold clamps can also be major sources of viruses. They are transmitted from these sources by aphids in the spring in a semi-persistent or persistent manner, depending on individual viruses. For example, BWYV is transmitted persistently for up to 50 days after a minimum feeding period of 5 minutes and a latent period of 12–24 hours. The disease is more severe after mild winters and warm springs, conditions conducive to aphid multiplication.

(v) *Significance*

Virus yellows is probably the most significant disease of sugar beet world-wide. It is very common, and yield losses of more than 50% have been reported.

(vi) *Control*

Cultural control. Overwintering sources of virus such as weeds and groundkeepers should be eliminated. Seed crops and fodder beet clamps should be kept away from root crops.

Chemical control methods are aimed at reducing aphid populations. Some products are incorporated into soil and others are foliar-applied. Aphicides approved for control of aphid pests in sugar beet include the following:

Active ingredient	Product example
aldicarb	Temik 10G
aldicarb + gamma-HCH	Sentry
carbosulfan	Marshal 10G

deltamethrin + heptenophos	Decisquick
demeton-S-methyl	Metasystox SS
dimethoate	Dimethoate
disulfoton	Campbell's Dusulfoton P10
fenvalerate	Sumicidin
oxamyl	Vydate
oxydemeton-methyl	Metasystox R
phorate	Phorate
pirimicarb	Aphox
thiometon	Ekatin

Strains of *Myzus persicae* have developed resistance to all
commercially available aphicides. Insecticides should, therefore,
only be used when necessary (see forecasting disease).

Forecasting disease. In the UK, forecasts for the amount of virus
yellows disease are made by Broom's Barn Experimental Station in
order to rationalise aphicide application. These regional forecasts
are based on previous years' virus yellows incidence in the area, the
number of viruliferous aphids caught in traps and the number of
ground frosts during winter.

Rhizomania

(i) *Causal organism*
Virus, beet necrotic yellow vein virus (BNYVV) spread by its
soil-borne fungal vector *Polymyxa betae.*

(ii) *Host range*
Sugar and fodder beet, spinach and several other Chenopodiaceae.

(iii) *Symptoms (Figure 4.3, p 71)*
Rhizomania (root madness) is often first noticed by foliar symptoms. Veins of infected plants first turn pale yellow, then brown and
leaves become elongated and erect. Such symptoms, however, are
not always apparent in the field. On roots, characteristic 'root
madness' symptoms occur. Proliferation of lateral rootlets results in
a bearding effect. The root also becomes constricted and vascular
tissue turns brown. Affected plants are extremely stunted and wilt
easily.

(iv) *Disease cycle*
The disease cycle of the virus is inherently associated with that of its
soil-borne fungal vector *Polymyxa betae.* The disease cycle of *P.*

betae is very similar to that already described for *P. graminis*, the vector of barley yellow mosaic virus.

(v) *Significance*

Rhizomania is a serious problem in most sugar beet growing areas in Western Europe. It is also an increasing problem in the USA. There was one outbreak of the disease in the UK in 1987 and despite efforts to contain the disease several further outbreaks occurred in 1989. Losses in sugar content of up to 70% have been reported as a result of severe attacks. The disease can clearly limit sugar beet production and is the most important disease of sugar beet where it occurs. It can survive in resting propagules of *P. betae* for at least 17 years.

(vi) *Control*

Legislative control attempts to exclude the disease from the UK. All beet species and seed potatoes entering the country must be from rhizomania-free areas. The disease is notifiable in the UK. In order to avoid the spread of contaminated soil and debris from imported potatoes and root vegetables, the *Disposal of Waste Order* was passed in 1988 in the UK (see Chapter 5, Legislative control).

Cultural control methods include the operation of wide rotations which can slow down the increase of disease. Well drained soil, together with careful use of irrigation and elimination of ground-keepers and weed beet, will also help. Disinfection of machinery and soil tillage along a single axis will reduce the spread of contaminated soil.

Disease resistance in beet varieties offers the best way of controlling the disease. Tolerant varieties are available in France but all varieties grown at present in the UK are susceptible. Much research using conventional and gene-insertion breeding techniques is under way aiming to produce rhizomania-resistant sugar beet.

Chemical control may be economic, but only on small patches of infested soil. Methyl bromide gas can be used as a soil sterilant, but it costs approximately £2000 per hectare and the treatment may need to be repeated every year.

Further reading

Benada, J., Sedivy, J. & Spacek, J. (1987). *Atlas of Diseases and Pests in Beet*. Amsterdam: Elsevier Science Publications.

Dunning, A. & Byford, W. (ed.) (1982). *Pests, Diseases and Disorders of Sugar Beet*. Distributed by Broom's Barn Experimental Station, Bury St Edmunds.

National Institute of Agricultural Botany (1989). Farmers' Leaflet No. 5, *Recommended Varieties of Sugar Beet*. Cambridge: NIAB publication.

Whitney, E. D. & Duffus, E. (ed.) (1986). *Compendium of Beet Diseases and Insects*. St Paul: The American Phytopathological Society.

11 Diseases of soybeans

Soybeans are an important world crop, grown mainly in the USA and South America, producing vegetable oil, high protein feed and soya protein.

In the USA, there has been an increase in the severity of a number of diseases in areas where continuous soybean production and reduced cultivation have been adopted.

Seedling blight

(i) *Causal organism*
Fungi, especially of genera *Alternaria*, *Fusarium*, *Pythium* and *Rhizoctonia*. Some pathogens which cause diseases of mature soybean plants such as charcoal rot (*Macrophomina phaseolina*) also cause seedling blight.

(ii) *Host range*
All the pathogens have a wide host range.

(iii) *Symptoms*
Seed-borne diseases may or may not be visible on seeds as blemishes and blotches. Badly contaminated seed or heavily contaminated soil can result in pre- and post-emergence damping-off (blight). Plants which survive may be stunted and unthrifty.

(iv) *Disease cycle*
The pathogens are seed and soil-borne. Seedling blight caused by *Fusarium* and *Pythium* spp. tends to be most severe in poorly drained wet soils. Blight caused by *Alternaria* and *Macrophomina* is more troublesome in dry sandy soils. Conditions which delay

emergence, such as cold weather and over-dense plant stands also encourage disease development.

(v) *Significance*

In a 1984 survey of loss estimates in the southern USA, seedling blight accounted for almost 1% of the losses (Mulrooney, 1986). Although crops can compensate for a certain amount of individual plant loss, severe seedling blight will reduce final yield significantly. Yield losses of up to 40% in individual crops have been recorded in the USA as a result of a damping-off and root rot, caused primarily by *Rhizoctonia solani*.

(vi) *Control*

Improvement in drainage and the use of fungicide seed treatments will reduce the problem.

Products approved for the control of damping-off or seedling blight include the following:

Active ingredient	*Product example*
carboxin	Vitavax (seed treatment)
mancozeb	Dithane
metalaxyl	Apron 25W, Ridomil 2E (soil applied)
quintozene (PCNB)	Terra-coat (seed treatment)

Phytophthora root and stem rot

(i) *Causal organism*

Fungus, *Phytophthora megasperma* f.sp. *glycinea* (*Phytophthora megasperma* var. *sojae*).

Twenty-four races of the pathogen have been so far identified.

(ii) *Host range*

Soybean, tomato, alfalfa (lucerne), clover, garden pea.

(iii) *Symptoms*

The fungus can cause pre- and post-emergence damping-off in seedlings. In mature susceptible plants, a brown girdling rot develops, followed by wilting and death. More resistant plants may survive an attack but brown sunken lesions may develop on stems. Diseased tap roots then may become covered with orange masses of spores of a secondary pathogen, *Fusarium* spp.

(iv) *Disease cycle*

The pathogen overwinters primarily as oospores in debris and soil. When temperatures rise and abundant moisture becomes available, oospores germinate to form sporangia which then release swimming zoospores. Zoospores are attracted to germinating seeds or rootlets where they encyst, germinate, and penetrate plant tissues. Secondary infections occur when sporangia, produced on rotting rootlets, release zoospores into flooded soils. Leaf and stem infections may also occur if contaminated soil is blown or splashed onto aerial plant parts and humid damp conditions prevail. The disease is most common in heavy, wet clay soils and on farms where minimum cultivation is practised. A temperature range between 25 and 30 °C favours disease development.

(v) *Significance*

The disease is widespread in almost all soybean growing areas. Yield losses may exceed 25% in severely affected, susceptible varieties.

(vi) *Control*

Cultural control. Direct improvement of drainage will reduce the disease. Ploughing soil also may indirectly improve drainage. Crop rotation is also advisable.

Disease resistance. Breeding for resistance to *P. megasperma* f.sp. *glycinea* was initiated in the mid-1950s and is race-specific. The long-term usefulness of such resistance is debatable because of the development of additional races which can overcome it. Emphasis is shifting towards breeding varieties with two or more genes for resistance and for varieties with tolerance rather than total resistance to the disease.

Chemical control. Seed treatments are available for the reduction of early attacks of *P. megasperma* f.sp. *glycinea*. Metalaxyl (Apron) and pyroxyfur (Grandstand) are systemic seed treatments with efficacy against the pathogen. Metalaxyl may be incorporated into soil or used as a foliar spray.

Biological control has been investigated and may prove feasible in the future. Hyperparasitic fungi on seed coats have been found to reduce oospore populations in soil (Filonow & Lockwood, 1985).

Integrated control offers the most rational approach to this disease. Two options were suggested by Schmitthenner (1985). The first is to combine the use of metalaxyl soil treatment with high-tolerant varieties; however, the cost may be prohibitive. The second is to combine high-tolerant varieties, ploughing, drainage, rotation, and metalaxyl seed treatment: the main problem with this approach is that some growers prefer not to plough fields.

Charcoal rot

(i) *Causal organism*
Fungus, *Macrophomina phaseolina*.

(ii) *Host range*
Extensive, including soybean and maize.

(iii) *Symptoms (Figure 11.1)*
Symptoms include damping-off of seedlings. In adult plants, after flowering, a silvery discoloration of the taproot and stem base develops. Vascular tissues become red-brown, the leaves chlorotic, and the plants wilt. Small black sclerotia, resembling small pieces of

Figure 11.1. Charcoal rot of soybeans caused by *Macrophomina phaseolina*. Note the presence of small black sclerotia below the epidermis. (Illinois Agricultural Experimental Station.)

charcoal develop just below the epidermis and give the disease its name.

(iv) *Disease cycle*
The fungus overwinters primarily as sclerotia in debris and soil, and as a contaminant of soybean seed coats and embryos. The fungus requires relatively high temperatures before sclerotia germinate and symptoms usually appear after flowering between 28 and 35 °C. Low soil moisture, dense plant populations, plant wounding and herbicide stress all encourage the disease.

(v) *Significance*
The disease occurs world-wide but tends to be more severe in areas where high soil temperature prevails, accompanied by low soil moisture. Plant losses in excess of 70% have been reported as a result of damping-off (*M. phaseolina*). In temperate areas the disease usually results in unthrifty rather than dead plants, but both yield and quality of the crop can be reduced.

(vi) *Control*
Cultural control methods include irrigation, crop rotation, optimum use of fertiliser and avoiding plant crowding.

Brown stem rot

(i) *Causal organism*
Fungus, *Phialophora gregata* (*Cephalosporium gregatum*).

(ii) *Host range*
Soybeans.

(iii) *Symptoms (Figure 11.2)*
The fungus causes a progressive necrosis of both the vascular and pith tissues at or after flowering. About 20–30 days before the plant reaches physiological maturity, a sudden interveinal chlorosis occurs, followed in some plants by wilting. The lower part of infected stems turn brown.

(iv) *Disease cycle*
The fungus overwinters in soil and plant debris. As temperatures rise, conidia are produced which may be wind and water dispersed in the crop. Mycelium in the soil may grow directly into roots,

colonise xylem vessels and spread up through the plant. Disease development is favoured between 15 and 27 °C but is inhibited over 32 °C. The relationship between disease severity and moisture is complex, but symptoms are evident at a period of intensive water uptake by the plant. The disease becomes progressively worse in continuous soybean crops grown with minimum cultivation.

(v) *Significance*

Yield losses are greatest if both stem and foliar symptoms occur. Losses of over 40% have been reported in individual, badly affected fields.

(vi) *Control*

The best methods of control are crop rotation and the use of disease-resistant varieties.

Figure 11.2. Brown stem rot of soybeans caused *Phialophora gregata*. Note brown vascular and pith tissues in split stems. (Illinois Agricultural Experiment Station.)

Pod and stem rot

(i) *Causal organism*
Fungus, *Diaporthe phaseolorum* var. *sojae* (*Phomopsis sojae*).

(ii) *Host range*
Mainly soybeans, but other crops and weeds may also be infected.

(iii) *Symptoms (Figure 11.3)*
All aerial parts of the soybean plant may be attacked, resulting in stunting and the production of numerous black pycnidia in bleached dead patches of leaf, stem and pod tissues. The disease on leaves usually starts at the margin and quickly progresses until the leaf is killed, rather than resulting in discrete lesions. Pycnidia develop in a linear fashion on stems and pods, in contrast to fruiting bodies produced in anthracnose which are scattered at random over affected tissue.

Figure 11.3. Pod and stem rot of soybeans caused by *Diaporthe phaseolorum* var. *sojae*. Note presence of numerous black pycnidia arranged in a linear fashion. (Illinois Agricultural Experiment Station.)

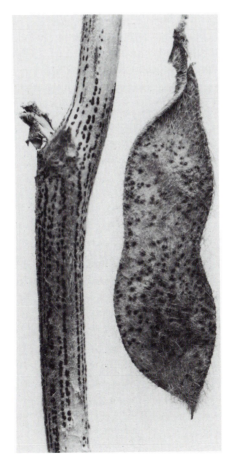

(iv) *Disease cycle*

The fungus overwinters as dormant mycelium in debris and seed. Primary infections are initiated by either asexual spores produced in pycnidia or sexual spores from perithecia. Free water is essential for spore germination. It is now known that the fungus infects at any time during the growing season and remains 'latent' or 'symptomless' until plants begin to senesce. It is at this stage that symptoms develop and fruiting structures are formed. However, if plants are stressed earlier in the season to cause early senescence, symptoms and fruiting structures will develop earlier. Temperatures above 20 °C together with prolonged wet periods tend to favour disease development. Secondary infections occur as a result of splash-dispersed asexual conidia. Colonisation of pods frequently leads to seed contamination particularly if harvest is delayed as a result of wet weather. Contaminated seed can result in the introduction of the disease into new planting areas.

(v) *Significance*

The disease is endemic in most soybean-growing areas of the world and caused the highest average loss (2%) of any soybean disease in a 1984 survey in southern USA (Mulrooney, 1986).

(vi) *Control*

Cultural control. Soybeans should be harvested promptly at maturity. Crop rotation and ploughing will also help to reduce the disease. High quality, pathogen-free seed should also be used.

Disease resistance is available in some soybean varieties. Varieties which mature during periods of low rainfall should also be selected.

Chemical control, using both seed treatments and foliar sprays, can be cost-effective on farms with a history of the problem (see also disease forecasting). Products approved for the control of pod and stem rot include the following:

Active ingredient	*Product example*
benomyl	Benlate
chlorothalonil	Bravo 500
thiophanate-methyl	Topsin

Biological control using bacterial seed treatments shows some promise (Cubeta, Hartman & Sinclair, 1985).

Forecasting disease, based on a points system, has been developed to rationalise fungicide usage. The system is based on cropping history, variety, planting date and rainfall. When a threshold level of points is reached, a fungicide spray should be applied.

Stem canker

(i) *Causal organism*
Fungus, *Diaporthe phaseolorum* var. *caulivora*.

(ii) *Host range*
Mainly soybeans, but other crops and weeds may also be infected.

(iii) *Symptoms*
Brick red lesions, usually on the nodes are the initial symptoms. Lesions then darken, elongate and deepen to become cankerous. Severe attacks result in interveinal chlorosis, necrosis of leaves and plant death.

(iv) *Disease cycle*
The disease cycle is similar to that described for *D. phaseolorum* var. *sojae*, although there is some debate as to the true significance of seed contamination.

(v) *Significance*
The disease has become of major significance since 1980 in southern USA, partly as a result of changes in agricultural practices and varieties. In a field where minimum cultivation was practised, an increase from 2% to 74% in plants showing symptoms was reported in one season. Yield losses of around 80% have been reported in susceptible varieties and financial losses of almost $40 million were estimated in soybean-producing states in the south-east USA during 1983 (Backman, Weaver & Morgan-Jones, 1985). A mathematical relationship between soybean yield and stem canker severity was given by Backman, Weaver & Morgan-Jones (1985).

(vi) *Control*
As for *D. phaseolorum* var. *sojae*.

Brown spot

(i) *Causal organism*
Fungus, *Septoria glycines*.

(ii) *Host range*
Soybeans.

(iii) *Symptoms*
Symptoms consist of irregular dark brown spots on soybean leaves. Lesions may coalesce; leaves become chlorotic and may drop off.

(iv) *Disease cycle*
The fungus overwinters mainly as pycnidia on debris and seed. During warm moist weather, pycnidia exude asexual conidia in a gelatinous matrix. Conidia are rain-splashed horizontally from plant to plant and vertically up plants. Colonisation of pods can lead to seed contamination. The disease tends to build up in continuous soybean cropping, especially in fields where minimum cultivation is practised.

(v) *Significance*
Yield losses of over 30% have been reported in individual crops. Yield reductions of 10% are common.

(vi) *Control*
Cultural control involves crop rotation, disposal of debris by ploughing, and planting pathogen-free seed.

Disease resistance is available in soybean varieties.

Chemical control may be worthwhile. Fungicides such as benomyl applied after flowering have increased yield by as much as 30%. Products approved for the control of brown spot are the same as those approved for control of pod and stem rot.

Anthracnose

(i) *Causal organism*
Fungi, *Colletotrichum truncatum* and *Glomerella glycines*.

(ii) *Host range*
Neither pathogen is specific to soybean and some weeds may be infected; *C. truncatum* also occurs on alfalfa and lima bean, but soybean is the primary host.

(iii) *Symptoms (Figure 11.4)*

The first symptom to be seen if contaminated seed is planted is damping-off. Dark sunken lesions, which develop on cotyledons of emerging seedlings, may cause defoliation and death. Symptoms of disease may or may not be seen on mature plant stems, pods, leaves and petioles. Irregular brown lesions develop and, particularly

Figure 11.4. Anthracnose of soybeans caused by *Colletotrichum truncatum* on a soybean stem (above) and pods (below). Note presence of numerous black acervuli arranged in a random fashion. (University of Illinois.)

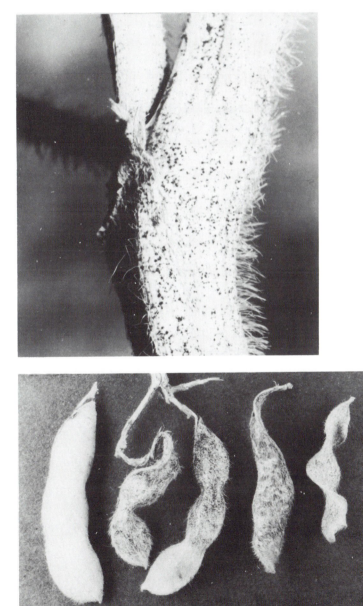

towards the end of the season, small black acervuli or perithecia develop in a random fashion on affected tissues (cf. pod and stem rot). Colonisation of pods can result in few small seeds and those remaining may become dark brown, shrivelled and mould-covered.

(iv) *Disease cycle*
Both pathogens overwinter as mycelium in debris, seed and weeds. Seed contamination can lead to damping-off or, alternatively, mycelium can become established in plants without symptoms until plants mature. Both pathogens produce asexual conidia in acervuli and *G. glycines* also produces sexual ascospores in perithecia. Wind and rain disperse spores. Relative humidities above 70% are necessary for infection and the disease is most serious in warm wet tropical or sub-tropical areas.

(v) *Significance*
The disease caused an average 1.1% yield loss, according to the 1984 survey of soybean diseases in southern USA (Mulrooney, 1986). Yield losses in individual fields have been reported in excess of 20%.

(vi) *Control*
Cultural control methods include crop rotation, correct disposal of debris by ploughing and use of high quality, pathogen-free seed.

Chemical control. Both fungicide seed treatments and foliar sprays are available. Benomyl has proved effective if applied between flowering and pod-fill. Products approved for the control of anthracnose are the same as those approved for the control of pod and stem rot.

Bacterial blight

(i) *Causal organism*
Bacterium, *Pseudomonas syringae* pv. *glycinea*.
  Ten races of the pathogen have been identified.

(ii) *Host range*
Soybeans.

(iii) *Symptoms*
First symptoms may be seen on cotyledons as dark brown necrotic lesions often starting from the cotyledon leaf margins. Lesions may

be covered with pale grey bacterial slime. On mature leaves, lesions are initially small, brown angular water-soaked spots. These enlarge and develop a dry centre, which may drop out, and a water-soaked margin. Lesions often coalesce and are frequently surrounded by a chlorotic halo. Severe attacks may defoliate lower branches. On stems large dark lesions develop, and on pods large water-soaked lesions similar to those on leaves occur. Contaminated seeds may become covered in a slimy bacterial growth.

(iv) *Disease cycle*

The pathogen overwinters on debris and seeds. Infection occurs through natural openings and wounds. Bacterial slime builds up in intercellular spaces and exudes onto plant surfaces. Secondary bacterial spread occurs by wind-driven rain. The disease is severe during periods of cool wet weather and is inhibited by hot dry weather.

(v) *Significance*

Bacterial blight is the most common bacterial disease on soybeans and estimated annual losses in excess of $60 million have been reported in the USA as a result of the disease (Kennedy & Alcorn, 1980).

(vi) *Control*

Cultural control. Crop rotation, ploughing and use of high quality pathogen-free seed will reduce losses due to the disease.

Disease resistance is available in soybean varieties and resistant varieties should be grown where the disease is a problem.

Virus diseases

According to Sinclair (1982), over 50 viruses or virus strains have been reported to cause diseases of soybeans throughout the world. Descriptions of bean pod mottle, Brazilian bud blight and yellow mosaic are given by Sinclair. Vectors of these three diseases are leaf beetles, aphids and aphids, respectively. Control strategies primarily involve the use of resistant varieties and elimination of potential weed hosts by herbicides.

Further reading

Sinclair, J. B. (ed.) (1982). *Compendium of Soybean Diseases*. St Paul: The American Phytopathological Society.

12 Diseases of maize (corn)

Maize is the third most important crop in the world after wheat and rice, providing a source of vegetable oil, high carbohydrate feed, maize meal and breakfast cereals. Largely a sub-tropical crop, maize is also grown in temperate areas, usually as a bulk fodder crop or for making silage.

Intensification of maize production in the USA has resulted in the upsurge of a number of diseases. Disease may affect maize stalks, leaves and ears. Storage moulds are also important, particularly if fungal contamination results in mycotoxin accumulation.

Stalk rot

(i) *Causal organism*
Fungi, *Diplodia maydis* (*Stenocarpella maydis*), *Fusarium culmorum*, *F. moniliforme* (*Gibberella moniliforme*) and *F. graminearum* (*G. zeae*). *Pythium aphanidermatum* and *Colletotrichum graminicola*, the cause of anthracnose, may also cause stalk rot.

(ii) *Host range*
The main stalk rot pathogens, the *Fusarium* spp. are most important on maize and other cereals. *Diplodia maydis* is largely confined to maize. *Pythium aphanidermatum* attacks a very wide host range and *C. graminicola* is confined mainly to cereals.

(iii) *Symptoms (Figure 12.1)*
Symptoms of *Diplodia* and *Fusarium* stalk rot are similar. Affected stalk tissue becomes spongy and bleached. Vascular bundles remain intact but pith tissue becomes discoloured. Dark sub-epidermal pycnidia of *Diplodia* can often be found near nodes, compared with

superficial perithecia formed in stalks affected by *Fusarium graminearum* (*Gibberella zeae*). In addition, *Fusarium*-rotted stalks often show a pinkish-red or orange discoloration, whereas a white fungal growth sometimes occurs in *Diplodia* stalk rot.

(iv) *Disease cycle*

Both *D. maydis* and the *Fusarium* species can overwinter on seed and crop debris. Seed contamination can result in damping-off. Generally stalk rot tends to be more severe if conditions are dry during the vegetative period of growth and wet during the reproductive period. Primary infections are initiated by wind-blown ascospores in *Fusarium graminearum* (*Gibberella zeae*), but subsequent infections by *Fusarium* spp. and infection by *D. maydis* arise mainly from rain-splashed conidia. Plants damaged by pest attack or adverse weather tend to be more susceptible to stalk rot.

Figure 12.1. Stalk rot of maize caused by *Fusarium graminearum* (*Gibberella zeae*). (University of Illinois.)

(v) *Significance*

Stalks rots are widespread and among the most destructive diseases of maize. The relationship between stalk rot and yield is controversial, but losses of up to 20%, as a result of poor grain-fill and lodging, are possible.

(vi) *Control*

Cultural control. This involves the use of a balanced fertiliser programme. Excess nitrogen should not be used and any deficiencies in potassium should be corrected. Optimum seed rates should be used and overcrowded humid crops avoided.

Disease resistance in varieties and hybrids is widely available.

Chemical control. Thiram and captan seed treatments will reduce seed contamination and damping-off, but do not protect plants from subsequent attacks.

Charcoal rot

(i) *Causal organism*

Fungus, *Macrophomina phaseolina*.

(ii) *Host range*

Very wide, including maize and soybeans.

(iii) *Symptoms*

The fungus can cause damping-off in maize, should heavily infested seeds be planted. Symptoms are subsequently seen typically as plants reach maturity. A silvery discoloration of stem base and root tissue develops. Brown water-soaked lesions then occur, which can result in premature ripening, shedding and lodging. Finally, numerous small grey-black sclerotia are formed, particularly along vascular bundles and in the pith cavity.

(iv) *Disease cycle*

The fungus overwinters on seed or as sclerotia in soil. Infection is more likely when plants are under stress; high temperatures (over 30 °C), low soil moisture, dense stands of plants and wounding all favour attacks.

(v) *Significance*
The disease is significant mainly as part of the stalk rot complex. It can build up on farms over years because of its wide host range.

(vi) *Control*
As for stalk rot. In addition, irrigation may reduce the effects of the disease, especially when applied at about the time of tasseling.

Stewart's disease (bacterial wilt)

(i) *Causal organism*
Bacterium, *Erwinia stewartii*.

(ii) *Host range*
Mainly maize and sweet corn.

(iii) *Symptoms (Figure 12.2)*
One of the most obvious symptoms is a rapid wilt of individual plants. Yellow, then brown streaks with irregular margins develop

Figure 12.2. Stewart's disease (bacterial wilt) of maize leaves caused by *Erwinia stewartii*.

in parallel lines on leaves. Tassels may also be affected; they become bleached and die prematurely. In severe attacks, cavities develop in the stalk near soil level and grains may also be contaminated.

(iv) *Disease cycle*
The main overwintering site is inside a hibernating insect vector, most commonly the corn flea beetle (*Chaetocnema pulicaria*). The vector transmits the bacterium from plant to plant during feeding, remaining infective for life. Plant nutrition appears to influence disease severity. High concentrations of nitrogen and phosphorus encourage disease, but plants containing high concentrations of calcium and potassium are more resistant. High temperatures also increase disease severity.

(v) *Significance*
The disease appears to be particularly severe in sweetcorn hybrids, where total crop losses have been reported.

(vi) *Control*
The use of resistant varieties and application of insecticides early in the growing season are the main control strategies. Disease forecasting, based on the sum of the mean temperatures for December, January and February may be useful in rational use of insecticides. Should this sum be less than 32 °C, most of the overwintering beetle vectors will have been killed and there is probably little necessity for insecticide applications. Temperature sums in excess of 37 °C indicate a high risk of disease later in the season.

Grey leaf spot

(i) *Causal organism*
Fungus, *Cercospora zeae-maydis* (*Cercospora sorghi* var. *maydis*).

(ii) *Host range*
Maize, some weeds and *Sorghum* spp.

(iii) *Symptoms*
Lower leaves are commonly affected initially, often rapidly becoming necrotic and dying. Lesions on mature higher leaves usually consist of angular pale brown or long rectangular grey patches. Severe attacks can result in defoliation and lodging.

(iv) *Disease cycle*

The fungus overwinters mainly as mycelium in surface debris. The survival of the fungus is greatly reduced if debris is buried. Long periods of high humidity, together with warm weather, encourage sporulation. Spores are usually rain-dispersed.

(v) *Significance*

The disease was relatively insignificant in the past because of the traditional agricultural practice of ploughing in surface debris. The recent trend towards minimum or no-till cultivation, together with continuous maize production, has resulted in an upsurge of the disease because of the increased inoculum on debris. Early attacks may result in a 20% yield loss.

(vi) *Control*

Cultural control by the use of more traditional practices such as ploughing and crop rotation would result in a rapid decline of the disease.

Disease resistance breeding is in progress and some resistant hybrids are now available.

Anthracnose

(i) *Causal organism*

Fungus, *Colletotrichum graminicola* (*Glomerella graminicola*).

(ii) *Host range*

Some small-grain cereals, grasses and maize.

(iii) *Symptoms*

Contaminated seed can lead to seedling blight. Symptoms on mature plants usually appear as small, round, water-soaked spots on leaves. These often enlarge and develop a dark brown centre with a red-orange border. In severe attacks, lesions coalesce, leaves become blighted, plant tops die back and a stalk rot may occur. Numerous black acervuli, which produce pink spores, may be visible in lesions.

(iv) *Disease cycle*

The fungus overwinters mainly as mycelium on debris although it can also be seed-borne. High temperatures and high humidity result in the production of conidia from acervuli, which are mainly rain-dispersed. Free water is required for infection to occur.

(v) *Significance*

Anthracnose is becoming a more significant disease, largely for the same reasons as grey leaf spot. Early attacks on susceptible varieties, usually as a result of warm wet weather, can result in significant yield losses.

(vi) *Control*

As grey leaf spot. Balanced soil fertility may also help.

Ear rot

(i) *Causal organism*

Several fungi cause ear rot. The most important are probably *Diplodia* and *Fusarium* spp. (*F. moniliforme, Gibberella zeae (F. graminearum)*) which have been described under stalk rots. However, Shurtleff (1980) describes a further seven fungal ear rots.

(ii) *Host range*

See stalk rot.

(iii) *Symptoms*

Diplodia ear rot symptoms consist of small, bleached greyish-brown ears with numerous black pycnidia. *Fusarium* ear rot symptoms consist of a reddish mould developing on the surface of ears. *Gibberella zeae* attacks can also result in the development of blue-black perithecia on husks.

(iv) *Disease cycle*

See stalk rots.

(v) *Significance*

Ear rots are more severe in humid areas where rainfall is above average during ear maturity and in crops which have suffered damage of some sort or have lodged. *Gibberella* ear rot sometimes reaches epidemic proportions in mid-western USA and losses in yield and quality of crop may be considerable. However, even when yield losses are minimised, *Fusarium* or *Gibberella* contamination of ears can result in the build-up in grain of mycotoxins which are potentially very harmful to livestock and humans.

(vi) *Control*

Early harvesting and the use of resistant varieties are the main methods of control.

Storage moulds

(i) *Causal organism*

Fungi, mainly of the genera *Aspergillus*, *Penicillium* and *Fusarium*.

(ii) *Host range*

Numerous.

(iii) *Symptoms*

Grey, green, blue, brown, black or pink mould development leading to heating and caking of seed.

(iv) *Disease cycle*

Affected seed may enter stores with or without disease symptoms as a result of ear rots in the field. Inoculum may also be present on debris in grain bins. Problems occur if seed has above 15% moisture content and if temperatures are allowed to exceed 21 °C.

(v) *Significance*

The major significance of storage moulds is the loss of grain and the possibility of mycotoxins building up in contaminated grain. Genera of all three types of storage mould can produce mycotoxins. *Aspergillus flavus* and related species produce aflatoxins which primarily damage liver tissue especially in younger animals. *Penicillium* species, including *P. expansum*, produce toxins which can also result in liver damage and cause haemorrhaging. Fusarium species, including *Gibberella zeae* (*F. graminearum*) and *F. moniliforme*, produce the toxins vomitoxin and zearalenone which cause vomiting and interfere with reproduction, respectively.

(vi) *Control*

Grain for storage should be dried to a minimum of 15% moisture (13% for long-term storage) as soon as possible after harvest. Grain should then be cooled to 2–5 °C and stored below 10 °C if possible. Grain bins should be thoroughly cleaned before use, ventilated throughout the storage period and checked regularly for evidence of heating, mustiness and mould development.

Virus diseases

More than 40 viruses have been reported as diseases of maize throughout the world and Shurtleff (1980) describes 12 of the most important ones in some detail. Virus symptoms can be extremely difficult to diagnose and frequently vary depending on factors such as plant age and variety. The main vectors of maize virus diseases are aphids, leafhoppers and beetles. Weeds should be eliminated and good crop rotations practised. There are varieties of maize which are resistant to certain virus diseases.

Further reading

Shurtleff, M. C. (ed.) (1980). *Compendium of Corn Diseases*. St Paul: The American Phytopathological Society.

13 Diseases of field vegetables

Field vegetables are specialist crops, generally regarded as horticultural crops. However, they are of increasing significance to arable farmers looking for alternatives to major arable crops in cases where financial returns have fallen in recent years. The most important field vegetables in the UK are brassicas (cabbages, cauliflowers and brussels sprouts), carrots, parsnips, onions, leeks and lettuces. As field vegetables are still considered to be minor crops, only a brief description of symptoms of the major diseases in each crop is shown, together with approved methods of chemical control. Fungicide recommendations are taken from *The UK Pesticide Guide* (Ivens, 1989), but it is always wise to consult manufacturers' product manuals before application. The use of fungicides on field vegetables increases as the pressure on farmers to produce high quality, blemish-free produce increases. It is particularly important to take note of manufacturer's recommendations regarding harvest intervals.

Vegetable brassicas

Most of the following vegetable brassica diseases have been described in detail for oilseed rape. The specified fungicides may not be approved for all vegetable brassicas.

Disease (causal organism)	Symptoms	Control
Clubroot (*Plasmodiophora brassicae*)	Wilt of leaves, swollen clubbed roots	ferbam + maneb + zineb (Trimanzone, soil applied), mercurous chloride (Calomel, soil applied), thiophanatemethyl (Mildothane, transplant-dip)

Disease (causal organism)	Symptoms	Control
Downy mildew (*Peronospora parasitica*)	Yellow angular patches with downy fungal growth underneath. Grey discoloration of cauliflower curds	chlorothalonil (Bravo 500), copper hydroxide (Chiltern Kocide 101), dichlofluanid (Elvaron), maneb, maneb + zinc (Manex), propamocarb hydrochloride (Fisons Filex, soil applied)
Powdery mildew (*Erysiphe cruciferarum*)	White fluffy mould, black flecks, chlorosis and necrosis	dinocap (Karathane Liquid), pyrazophos (Missile), triadimefon (Bayleton), triadimenol (Bayfidan)
Light leaf spot (*Pyrenopeziza brassicae*)	Bleached spots with white spore masses	benomyl (Benlate)
Phoma leaf spot (*Leptosphaeria maculans*)	Bleached spots with black pycnidia	—
Dark leaf spot (*Alternaria*) (*Alternaria brassicae* and *A. brassicicola*)	Target spots, dark brown lesions on cauliflower curds	iprodione (Rovral Flo), maneb, maneb + zinc (manex)
Ring spot (*Mycosphaerella brassicicola*)	Target spots made up of black pycnidia	benomyl (Benlate), chlorothalonil (Bravo 500)
White blister (*Albugo candida*)	White raised powdery patches	mancozeb + metalaxyl (Fubol 58WP) (Brussels sprouts), chlorothalonil + metalaxyl (Folio 575FW) (Brussels Sprouts)
Botrytis grey mould (*Botrytis cinerea*)	Grey mould on dead/ decaying tissue	chlorothalonil (Bravo 500), iprodione (Rovral Flo)

Carrots and parsnips

Disease (causal organism)	Symptoms	Control
Alternaria blight and black rot (*Alternaria dauci* and *A. radicina*)	Rapidly spreading brown lesions with chlorotic margins on leaves (blight), dark rot of	copper hydroxide (Chiltern Kocide 101) (blight), benomyl (Benlate as dip prior

Disease (causal organism)	Symptoms	Control
Carrots only	carrot roots especially in store (black rot).	to storage) (black rot), **iprodione + metalaxyl + thiabendazole (Polycote Prime)** (seed treatment–carrots)
Powdery mildew (*Erysiphe heraclei* in carrots, *E. polygoni* in parsnips)	White fluffy mould	triadimefon (Bayleton)
Canker (*Itersonilia pastinacae*) **Parsnips only**	Deep dark lesions especially on shoulder of roots	—
Cavity spot (*Pythium* spp.) Carrots only	Small clear or dark oval depressions on roots	mancozeb + metalaxyl (Fubol 58WP)
Scab (*Streptomyces scabies*) Carrots only	Raised corky lesions on roots	—
Sclerotinia rot (*Sclerotinia sclerotiorum*)	Foliage dies from base, white mould and black sclerotia develop. Can develop in store	benomyl (Benlate as dip prior to storage)
Violet root rot (*Helicobasidium purpureum*) Carrots only	Foliage rot, felt-like mass, purple fungal strands on roots	—

Storage diseases of carrots, including grey mould caused by *Botrytis cinerea* and various other rots, can be reduced by benomyl (Benlate) pre-storage dip, but fungicide-resistant strains of *B. cinerea* are common.

Onions and leeks

Disease (causal organism)	Symptoms	Control
Downy mildew (*Peronospora destructor*)	Pale oval lesions on leaves, grey fungal growth. Leaf-tip dieback	chlorothalonil + metalaxyl (Folio 575FW), copper hydroxide (Chiltern Kocide 101), ferbam + maneb + zineb (Trimanzone), propamocarb hydrochloride (Fisons Filex–soil applied), zineb

Disease (causal organism)	Symptoms	Control
Leaf rot (*Botrytis squamosa and B. cinerea*) Onions only	Small circular white spots, water-soaked margin. Leaf-tip dieback	benomyl (Benlate) carbendazim (Bavistin), iprodione (Rovral Flo), vinclozolin (Ronilan)
Leaf blotch (*Cladosporium allii-cepae* on onions, *C. allii* on leeks)	White spots, darken with age	chlorothalonil (Bombadier)
White tip (*Phytophthora porri*)	Bleached leaf tips	ferbam + maneb + zineb (Trimanzone), metalaxyl + chlorothalonil (Folio 575FW), propamocarb hydrochloride (Fisons Filex)
Rust (*Puccinia porri*) Leeks only	Elongated orange pustules on leaves	fenpropimorph (Corbel), ferbam + maneb + zineb (Trimanzone), triadimefon (Bayleton)
Smut (*Urocystis cepulae*)	Lead-coloured stripes in leaves. Black smut spores	—
White rot (*Sclerotium cepivorum*)	Yellow leaves, rotten roots, white fungal growth on stem base with black sclerotia	bromophos + captan + thiabendazole (Bromotex T) (seed treatment), mercurous chloride (Calomel, soil applied)
Fusarium foot rot (*Fusarium culmorum*) Leeks only	Red discoloration at stem base. Wilt and rot of leaves	—
Neck rot (*Botrytis allii*) Onions only	Soft bulbs, brownish rot of scales. Storage disease	benomyl (Benlate, seed treatment), benomyl + iodofenphos + metalaxyl (Polycote Pedigree, seed treatment), bromophos + captan + thiabendazole (Bromotex T, seed treatment), iprodione (Rovral Flo), thiabendazole + thiram (Hy-Vic, seed treatment)

Lettuces

Disease (causal organism)	Symptoms	Control
Downy mildew (*Bremia lactucae*)	Yellow patches on leaves with grey downy fungal growth underneath	mancozeb (Karamate N), zineb, mancozeb + metalaxyl (Fubol 58 WP), metalaxyl + thiram (Favour 600FW), zineb (Hortag)
Ring spot (*Marssonina panattoniana*)	Circular brown spots, shot-hole effect	—
Grey mould (*Botrytis cinerea*)	Grey mould on decaying tissue, plant collapse	benomyl (Benlate), iprodione (Rovral Dust), quintozene (Brabant PCNB 20%, soil applied), thiram (Hortag), vinclozolin (Ronilan)
Sclerotinia rot (*Sclerotinia sclerotiorum*)	White fluffy mould on affected tissue with sclerotia, plant collapse	quintozene (Brabant PCNB 20%, soil applied)
Bottom rot (*Rhizoctonia solani*)	Reddish-brown lesions on leaf bases, brown fungal threads	quintozene (Brabant PCNB 20%, soil applied), tolclofos-methyl (Basilex, soil applied)
Bacterial wilt (*Pseudomonas* and *Erwinia* spp.)	Red-brown vascular tissue, stem cavities, slimy green rot, wilting	—

Virus diseases of lettuce include lettuce mosaic, cucumber mosaic, beet western yellows and big vein. Vectors for all these virus diseases except big vein are aphids, which may be controlled by use of aphicides. Big vein is spread by the soil-borne fungus *Olpidium brassicae* and can be controlled by carbendazim (Bavistin) incorporated into peat blocks.

Further reading

Association of Applied Biologists (1986). Wellesbourne, Warwick: *Aspects of Applied Biology 12, Crop Protection in Vegetables*.

Dixon, G. R. (1981). *Vegetable Crop Diseases*. London: Macmillan.

Dunning, R. A., Byford, W., Dunn, J. A., Gair, R., Humphreys-Jones, R. & Maude, R. B. (1980). *Pest and Disease Control in Vegetables, Potatoes and Sugar Beet*. Thornton Heath, Surrey: BCPC publication.

Ministry of Agriculture, Fisheries and Food (1982). *Control of Pests and Diseases of Field Vegetables*. ADAS Booklet 2383. Alnwick, Northumberland: MAFF Publications.

Sherf, A. F. & Macnab, A. A. (1986). *Vegetable Diseases and their Control*. New York: John Wiley.

Glossary of plant pathological terms

Acervulus (pl. **acervuli**) A subepidermal, cushion-like mass of hyphae containing asexual conidia and conidiophores.

Active ingredient The active component of a formulated product.

Adult plant resistance Resistance detectable at the post-seedling stages of development (mature plant resistance). See also field resistance.

Aeciospore An asexually produced dikaryotic rust spore found in an aecium.

Aecium (pl. **aecia**) A cup-shaped fruiting body of rust fungi in which aeciospores are borne.

Aerobic Requiring free oxygen for respiration.

Aflatoxins Mycotoxins produced by *Aspergillus flavus*.

Alternate host One of two hosts required by a pathogen to complete its life-cycle.

Alternative host One of several plant species hosts of a given pathogen.

Anaerobic Living in the absence of free oxygen.

Antagonist An organism used in biological control which can decrease the capacity of a pathogen to cause disease.

Anthracnose A plant disease having characteristic limited lesions, necrosis and hypoplasia.

Apoplast Non-living parts of the plant including cell walls and xylem.

Apparent infection rate Rate of infection, r, calculated in Vanderplank's model of disease development, which takes into account the amount of plant tissue left to be colonised.

Appressorium (pl. **appressoria**) A swollen fungal hyphal tip usually associated with the mechanism of adherence to the plant surface prior to penetration.

Ascocarp Fruiting body in or on which asci are produced.

Ascospore A sexually produced spore borne in an ascus.

Ascus (pl. **asci**) A spore sac containing ascospores.

Asexual Vegetative.

Autoecious Completing the life-cycle on one host (especially rusts). See heteroecious.

Avirulent Lacking virulence, non-pathogenic.

Basidiocarp Fruiting body in or on which basidia are produced.

Basidiospore A sexually produced spore borne on a basidium.

Basidium (pl. **basidia**) A club-shaped structure on which sexually produced basidiospores are borne. Sometimes called promycelium.

Biological control The reduction of the amount of inoculum or disease-producing activity of a pathogen accomplished by or through one or more organisms (antagonists) other than man.

Biotroph An organism which is entirely dependent upon another living organism as a source of nutrients (obligate parasite).

Binary fission Simple asexual division of a cell, especially bacteria.

Blight A disease characterised by the rapid death of plant tissue.

Boom and bust cycle Situation, particularly with regard to cereal varieties when, after a period of widespread cultivation of a variety with major gene resistance (boom), it succumbs to disease, and its popularity declines (bust).

Broad-spectrum fungicide A fungicide with activity against a wide range of pathogens.

Canker A necrotic often sunken lesion.

Capsule A thick well developed slime layer surrounding a bacterium.

Chemotherapy The treatment of disease by chemical means.

Chlamydospore An asexually produced, thick-walled resting spore.

Chlorosis Yellowing of usually green plant tissue.

Circulative virus A virus which passes through the gut wall of the vector into the haemolymph and eventually contaminates the mouthparts via the saliva.

Cirrus (pl. **cirri**) A gelatinous mass of extruded spores.

Cleistothecium (pl. **cleistothecia**) A closed, often spherical ascocarp.

Coenocytic A continuous mass of cytoplasm and nuclei unbroken by septa.

Colonisation The spread of the pathogen in the host tissue away from the initial site of infection and the dependence on the host for nutrients.

Compound-interest disease A disease which goes through more than one cycle of infection during a growing season, analagous to a bank account giving compound interest.

Conidiophore A specialised hyphal branch bearing conidia.

Conidium (pl. **conidia**) An asexually produced fungal spore.

Cull pile A discarded pile of plant material, especially potato tubers.

Cyst A sac, especially a resting spore or sporangium-like structure.

Damping-off The rot of seedlings near soil level after emergence (post-emergence) or before emergence (pre-emergence).

Differential variety A variety which gives reactions which distinguish between race-specific isolates of a pathogen.

Dikaryotic Containing two sexually compatible nuclei per cell.

Diploid Having two sets of chromosomes.

Disease A harmful deviation from normal functioning of physiological processes.

Disinfestation The destruction of a pathogen on the surface of the host or in the environment surrounding the host.

Durable resistance Resistance which remains effective to pathogens which have highly developed variety-specific pathogenicity even though varieties are extensively cultivated in environments favourable to disease.

Enation Tissue malformation or gall production often induced by virus infection.

Epidemic A progressive increase in the incidence of a particular disease within a defined host population.

Epidemiology The study of factors influencing the development of a disease epidemic.

Eradicant fungicide A fungicide used to kill existing pathogen infestation. Often referred to as a curative fungicide.

Ergosterol A major fungal sterol.

Exudate Substance passed from within a plant to the outer surface or into the surrounding medium.

Facultative parasite An organism able to live as a saprophyte or a parasite.

Fasciation Striped pattern.

Field resistance Resistance detectable under natural infection in field conditions. See adult plant resistance.

Flagellum (pl. **flagella**) A whip-like organ of motility found on bacteria and zoospores.

Forma specialis (pl. **formae speciales**) A sub-division of an organism characterised from a physiological standpoint (especially host adaptation).

Fumigation Disinfestation by fumes.

Fungicidal Able to kill fungal spores or mycelium.

Fungicide A substance that kills fungal spores or mycelium.

Fungicide resistance A decrease in sensitivity to a fungicide due to selection or mutation following exposure to the compound.

Fungistatic Able to stop fungal growth without killing the fungus.

Gall An abnormal growth or swelling produced as a result of pathogen invasion.

Gene-for-gene hypothesis The concept that corresponding genes for resistance and virulence exist in host and pathogen, respectively.

Germ tube The initial hyphal growth from a germinating fungal spore.

Green bridge Living plant material used by biotrophs to overwinter.

Green island effect A green ring of tissue in a senescent leaf surrounding an individual rust pustule.

Groundkeeper A self-sown plant, especially potato.

Haploid Having one set of chromosomes.

Haulm Above-ground parts of a plant (stem and leaves), especially of potatoes.

Haustorium (pl. **haustoria**) A specially developed fungal hyphal branch within a living cell of the host for absorption of food.

Heteroecious Requiring two host species to complete its life-cycle. Especially in rust fungi.

Heterothallic The condition in which sexual reproduction can only occur between different sexually compatible mycelia (thalli).

Homothallic The condition in which sexual reproduction can occur within the fungal mycelium (thallus).

Horizontal resistance Resistance which is evenly spread against all races of a pathogen.

Host An organism harbouring a parasite.

Hybridisation The crossing of two individuals differing in one or more heritable characteristics resulting in the production of a hybrid.

Hydrophilic Affinity for water.

Hyperparasitism Parasitism on another parasite.

Hyperplasia Abnormal growth associated with increased cell division.

Hypersensitivity A rapid local reaction of plant tissue to attack by a pathogen resulting in the death of tissue around infection sites preventing further spread of infection.

Hypertrophy Abnormal growth associated with cell enlargement.

Hypha (pl. **hyphae**) A tubular thread-like filament of fungal mycelium.

Hyphopodium (pl. **hyphopodia**) A short mycelial branch.

Immune Exempt from infection.

Imperfect state The asexual period of a fungal life-cycle.

Incubation period The period of time between infection and the appearance of symptoms.

Infection The penetration of the host by a pathogen and the earliest stages of development within the host.

Infection court The initial site of contact between a pathogen and the surface of the host.

Infection peg A slender structure formed by the deposition of substances such as lignin around a thin hypha penetrating a host cell.

Inoculum (pl. **inocula**) Spores or other pathogen parts which can cause disease.

Integrated control The continuous application of a balanced range of disease control measures.

Intercellular Between cells.

Intracellular Within or through cells.

Koch's postulates Criteria proposed by Koch for proving the pathogenicity of an organism.

Latent period The time between infection and sporulation of the pathogen on the host, or time from the start of a virus vector's feeding period until the vector is able to transmit the virus to healthy plants.

Lesion A localised area of diseased or disordered tissue.

Lignituber A structure formed by the deposition of lignin surrounding the tip of a fungal hypha penetrating a host cell. Assumed to function as an active resistance mechanism.

Lipophilic Affinity for fat.

Lodging Breakage of many plant stems, especially cereals resulting in many tillers falling down.

Major gene resistance Genetic resistance to disease based on one or a few genes.

Monocyclic Having only one cycle of infection during a growing season (see simple-interest disease).

Mosaic Patchy variation of normal green colour. Symptomatic of many virus diseases.

Mottle An arrangement of indistinct light and dark areas. Symptomatic of many virus diseases.

Multiline A combination of almost genetically identical breeding lines (isogenic) which have all agronomic characters in common, but differ in major gene resistance.

Mycelium A mass of hyphae that form the vegetative body of a fungus.

Mycotoxin Toxins produced by fungi which may contaminate foodstuffs.

Necrosis A browning or blackening of cells as they die.

Necrotroph An organism that causes the death of host tissues as it grows through them such that it is always colonising dead substrate.

Non-persistent virus A virus that persists in its vector for a few (usually less than 4) hours at approximately 20 °C.

Notifiable disease A disease which by law has to be reported to the appropriate authorities.

Obligate parasite An organism capable of living only as a parasite.

Oospore A sexually produced resting spore of the oomycetes.

Pathogen An organism which causes disease.

Pathovar (Pathotype) A subdivision of a species distinguished by common characters of pathogenicity, particularly in relation to host range.

Parasite An organism or virus which lives on another living organism (host), obtaining its nutrient supply from the host but conferring no benefit in return.

Perfect state Stage in the life-cycle of a fungus characterised by sexual spores.

Perithecium (pl. **perithecia**) A closed flask-shaped ascocarp having an apical hole.

Peritrichous Having a number of flagellae over the surface of a bacterium.

Persistence Time for which a virus vector remains infective after leaving the virus source.

Persistent virus A virus which persists in vector for more than 100 hours and in some cases for the life of the vector.

Phage (Bacteriophage) A virus which attacks bacteria.

Phyllody The replacement of floral parts by leaf-like structures.

Phytoalexin A substance which inhibits the development of a micro-organism, produced in higher plants in response to certain stimuli (biological, chemical or physical).

Phytosanitary certificate A certificate of health which accompanies plants or plant products to be exported.

Phytotoxic Toxic to plants.

Plasmid A self-replicating extra-chromosomal circle of DNA.

Plasmodium (pl. **plasmodia**) A naked amoeboid multinucleate mass of protoplasm.

Polycyclic Having more than one cycle of infection during a growing season (see compound-interest disease).

Polygenic resistance Genetic resistance to disease based on many genes.

Propagative virus A virus which multiplies in its vector.

Propagule That part of an organism by which it may be dispersed or reproduced.

Prophylaxis Preventative treatment against disease.

Protectant fungicide A fungicide which protects against invasion by a pathogen.

Pseudothecium (pl. **pseudothecia**) A fruiting body containing asci similar in appearance to a perithecium, but produced in an aggregation of vegetative hyphae.

Pustule A blister-like spore mass breaking through a plant epidermis.

Pycnidium (pl. **pycnidia**) A flask-shaped or spherical fungal receptacle bearing asexual spores, pycnospores.

Pycnospore An asexual spore produced in a pycnidium.

Quarantine The holding of imported material in isolation for a period to ensure freedom from diseases and pests.

Race A genetically and often geographically distinct mating group within a species. Also a group of pathogens distinguished by their ability to infect a given set of plant varieties.

Race non-specific resistance Resistance to all races of a pathogen.

Race-specific resistance Resistance to some races of a pathogen, but not to others.

Resistance factor Resistance gene or genes in a host which have not necessarily been identified but can be used for practical purposes in gene-for-gene relationships (see virulence factor).

Resistant Possessing qualities which prevent or retard the development of a given pathogen.

Resting spore A thick-walled spore that remains dormant for a period of time before germination.

Rhizosphere The zone in soil affected by roots.

Roguing Removal of diseased or unwanted plants from a crop.

Rohr A sock-like structure facilitating host penetration by *Plasmodiophora brassicae*.

Rot Disintegration of tissue.

Saprophyte An organism that lives on dead and decaying material.

Scab A roughened incrustation. A disease in which such lesions form.

Sclerotium (pl. **sclerotia**) A long-lived compacted mass of vegetatively produced hyphae.

Seedling resistance Resistance detectable at the seedling stage.

Semi-persistent virus A virus which persists in its vector for between 10 and 100 hours.

Septum (pl. **septa**) A partition or cross-wall.

Seta (pl. **setae**) A sterile hair.

Simple-interest disease A disease which goes through only one cycle of infection during a growing season analogous to a bank account giving simple interest.

Smith period (**Blight infection period**) Two consecutive days (ending 0900 hours) when the temperature has not been less than 10 °C and relative humidity above 90 % for at least 11 hours of each day. Used in forecasting outbreaks of late blight of potatoes (*Phytophthora infestans*).

Spermatium (pl. **spermatia**) A gamete produced in a spermogonium.

Spermogonium (pl. **spermogonia**) A fruiting body in which gametes (spermatia) are produced.

Sporangiophore A specialised hyphal branch bearing sporangia.

Sporangiospore A non-motile asexual spore produced in a sporangium.

Sporangium (pl. **sporangia**) A container which produces asexual spores. Sometimes functions as a single spore.

Spore A specialised propagative or reproductive body in fungi.

Spraing Scottish dialect word for streaks or stripes; a virus disease of potatoes.

Spreader A substance added to a spray to assist in its even distribution over the target.

Stachel A bullet-shaped structure used by *Plasmodiophora brassicae* to directly penetrate a host cell.

Sterilisation The elimination of micro-organisms.

Sticker A substance added to a spray to assist in its adhesion to the target.

Straggling Breakage of a few plant stems, especially cereals resulting in a few tillers falling down.

Stroma (pl. **stromata**) A mass of vegetative hyphae in or on which spores are produced.

Stylet-borne virus A virus which is borne on the stylet of its vector.

Super-race A pathogen race which contains virulence factors to match any resistance factors available in the host.

Suppressive soil Soil in which a pathogen may persist, but either causes little or no damage or causes disease for a short time and then declines.

Surfactant A surface active material, especially a wetter or spreader used with a spray.

Susceptible Subject to infection.

Symbiosis A mutually beneficial association of two or more different kinds of organisms.

Symplast Living parts of the plant including the phloem and protoplast.

Symptom A visible change in a host plant as a result of pathogen infection.

Systemic fungicide A fungicide which is absorbed and translocated in the plant.

Systemic infection Infection throughout the plant.

Take-all decline The decline in the cereal disease take-all after three or four successive cereal crops.

Target spot A lesion consisting of a dark brown circular area containing brown concentric rings. Typical of infection by *Alternaria* spp.

Teliospore A resting spore of rust and smut fungi in which sexual fusion occurs.

Tolerant Able to endure infection by a pathogen without showing severe symptoms of disease, or, able to compensate for the effects of disease.

Translaminar activity Ability of a fungicide to move through a leaf from one surface to the other.

Tylose A balloon-like intrusion into the lumen of a xylem vessel, formed by enlargement of an adjacent living cell and its extrusion through a pit in the vessel wall.

Tylosis The process of tylose formation.

Uredium (pl. **uredia**) A fruiting body of the rust fungi in which uredospores are produced.

Uredospore An asexual spore of the rust fungi.

Vector An organism which transmits a pathogen, usually a virus.

Vertical resistance Resistance to some races of a pathogen but not to others.

Vertifolia effect Proposed by Vanderplank as the erosion of horizontal resistance when plant breeding material is protected from disease by vertical resistance.

Virulence The degree or measure of pathogenicity.

Virulence factor Virulence gene or genes in a pathogen which have not necessarily been identified but can be used for practical purposes in gene-for-gene relationships (see resistance factor).

Viruliferous A vector which carries and can transmit a virus.

Volunteer plant A self-sown plant, especially cereal.

Whitehead A bleached cereal ear containing little or no grain. Usually a result of attack by stem base or root pathogens, particularly *Gaeumannomyces graminis* (take-all).

Wilt Loss of turgor in plant parts resulting in drooping.

Witches' broom Abnormal proliferation of shoots.

Yellows A plant disease characterised by a yellowing of tissue.

Zoosporangium (pl. **zoosporangia**) A sporangium containing or producing zoospores.

Zoospore A fungal spore capable of movement in water.

Zygospore A sexually produced resting spore of the zygomycetes.

Text references

Agrios, G.N. (1978). *Plant Pathology*, 2nd edn, p. 5. New York: Academic Press.

Agrios, G. N. (1988a). *Plant Pathology*, 3rd edn, p. 21. New York: Academic Press.

Agrios, G.N. (1988b). *Plant Pathology*, 3rd edn, pp. 516–17. New York: Academic Press.

Ainsworth, G. C. (1981). *Introduction to the History of Plant Pathology*. Cambridge: Cambridge University Press.

Anon. (1947). The measurement of potato blight. *Transactions of the British Mycological Society*, **31**, 140–1.

Asher, M. J. C. (1986). The occurrence and control of sugar beet powdery mildew. In *Aspects of Applied Biology 13, Crop Protection of Sugar Beet and Crop Protection and Quality of Potatoes*, pp. 131–7. Wellesbourne, Warwick: Association of Applied Biologists.

Ashworth, L. J., Jr., Huisman, O. C., Harper, D. M., Stromberg, L. K., & Bassett, D. M. (1979). Verticillium wilt disease of cotton: Influence of inoculum density in the field. *Phytopathology*, **69**, 483–9.

Backman, P. A., Weaver, D. B. & Morgan-Jones, G. (1985). Soybean stem canker: an emerging disease problem. *Plant Disease*, **69**, 641–7.

Baker, R. T. (1985). Biological control of plant pathogens: Definitions. In *Biological Control in Agricultural IPM Systems*, ed. M. A. Hoy & D. C. Herzog, pp. 25–39. Florida: Academic Press.

Ballinger, D. J. & Kollmorgen, J. F. (1986). Control of take-all of wheat in the field with benzimidazole and triazole fungicides applied at seeding. *Plant Pathology*, **35**, 67–73.

Bateman, G. L. (1984). Soil-applied fungicides for controlling take-all in field experiments with winter wheat. *Annals of Applied Biology*, **104**, 459–65.

Bradbury, J. F. (1986). *Guide to Plant Pathogenic Bacteria*. Slough: Commonwealth Agricultural Bureaux.

Brenchley, G. H. & Wilcox, H. J. (1979). *Potato Diseases*, pp. 26–7. ADAS RPD1, MAFF publication. London: HMSO.

Brent, K. (1985). One hundred years of fungicide use. In *Fungicides for Crop Protection*, BCPC Monograph 31, ed. I. M. Smith, pp. 11–22. Thornton Heath, Surrey: BCPC publication.

Clarkson, J. D. S. (1981). Relationship between eyespot severity and yield loss in winter wheat. *Plant Pathology,* **30**, 125–31.

Clarkson, J. D. S. & Cook, R. J. (1983). Effect of sharp eyespot (*Rhizoctonia cerealis*) on yield loss in winter wheat and of some agronomic factors on disease incidence. *Plant Pathology,* **32**, 421–8.

Cockbain, A. J. (1983). Viruses and virus-like diseases of *Vicia faba* L. In *The Faba Bean (Vicia faba* L.), ed. P. D. Hebblethwaite, pp. 421–62. London: Butterworths.

Cook, R. J. & Baker, K. F. (1983a). *The Nature and Practice of Biological Control of Plant Pathogens*, p. 60. St Paul: The American Phytopathological Society.

Cook, R. J. & Baker, K. F. (1983b). *The Nature and Practice of Biological Control of Plant Pathogens*, pp. 312–89. St Paul: The American Phytopathological Society.

Cook, R. J. & King, J. E. (1984). Loss caused by cereal diseases and the economics of fungicidal control. In *Plant Diseases, Infection, Damage and Loss*, ed. R. K. S. Wood & G. J. Jellis, p. 238. Oxford: Blackwell.

Cubeta, M. A., Hartman, G. L. & Sinclair, J. B. (1985). Interaction between *Bacillus subtilis* and fungi associated with soybean seeds. *Plant Disease,* **69**, 506–9.

Dekker, J. (1984). Development of resistance to antifungal agents. In *Mode of Action of Antifungal Agents*, ed. A. P. J. Trinci & J. F. Ryley, pp. 89–111. Cambridge: Cambridge University Press.

Dickinson, C. H. & Lucas, J. A. (1982a). *Plant Pathology and Plant Pathogens*. Basic Microbiology, Vol. 6, ed. J. F. Wilkinson, p. 94. Oxford: Blackwell.

Dickinson, C. H. & Lucas, J. A. (1982b). *Plant Pathology and Plant Pathogens*. Basic Microbiology, Vol. 6, ed. J. F. Wilkinson, p. 170. Oxford: Blackwell.

Farrar, J. F. (1984). Effects of pathogens on transport systems. In *Plant Diseases, Infection Damage and Loss*, ed. R. K. S. Wood & G. J. Jellis, pp. 87–104. Oxford: Blackwell.

Filonow, A. B. & Lockwood, J. L. (1985). Evaluation of several Actinomycetes and the fungus *Hyphochytrium catenoides* as biocontrol agents for Phytophthora root rot of soybean. *Plant Disease,* **69**, 1033–6.

Fitt, B. D. L. & Lysandrou, M. (1984). Studies on mechanisms of splash dispersal of spores, using *Pseudocercosporella herpotrichoides* spores. *Phytopathologische Zeitschrift,* **111**, 323–31.

Fry, W. E. (1987). Advances in disease forecasting. In *Rational Pesticide Use*, ed. K. J. Brent & R. K. Atkin, pp. 239–52. Cambridge: Cambridge University Press.

Garrett, S. D. (1970). *Pathogenic Root-Infecting Fungi*. Cambridge: Cambridge University Press.

Gibbs, A. & Harrison, B. D. (1976). *Plant Virology: the Principles*. London: Arnold.

Habeshaw, D. (1984). Effects of pathogens on photosynthesis. In *Plant Diseases, Infection, Damage and Loss*, ed. R. K. S. Wood & G. J. Jellis, pp. 63–72. Oxford: Blackwell.

Hagedorn, D. J. (ed.) (1984). *Compendium of Pea Diseases*. St Paul: American Phytopathological Society.

Hagedorn, D. J. (1985). Diseases of peas: their importance and opportunities for breeding for disease resistance. In *The Pea Crop*, ed. P. D. Hebblethwaite, M. C. Heath & T. C. K. Dawkins, pp. 205–13. London: Butterworths.

Hawksworth, D. L., Sutton, B. C. & Ainsworth, G. C. (1983). *Dictionary of the Fungi*. Slough: Commonwealth Agricultural Bureaux.

Hayes, J. D. & Johnston, T. D. (1971). Breeding for disease resistance. In *Diseases of Crop Plants*, ed. J. H. Western, pp. 62–88. London: Macmillan.

Henis, Y. Ghaffar, A. & Baker, R. (1978). Integrated control of *Rhizoctonia solani* damping-off of radish: effect of successive plantings, PCNB and *Trichoderma harzianum* on pathogen and disease. *Phytopathology,* **68**, 900–7.

Hims, M. J. (1987). The effects of disease control on quality of winter barley. In *Aspects of Applied Biology 15, Cereal Quality*, pp. 431–8. Wellesbourne, Warwick: Association of Applied Biologists.

Hirst, J. M., Stedman, D. J. & Hurst, G. W. (1967). Long distance spore transport: vertical sections of spore clouds over the sea. *Journal of General Microbiology,* **33**, 335–44.

Hollings, M. (1983a). Virus diseases. In *Plant Pathologist's Pocketbook*, ed. A. Johnston & C. Booth, p. 46. Slough: Commonwealth Agricultural Bureaux.

Hollings, M. (1983b). Virus diseases. In *Plant Pathologist's Pocketbook*, ed. A. Johnston & C. Booth. pp. 60–61. Slough: Commonwealth Agricultural Bureaux.

Hollins, T. W., Lockley, K. D., Blackman, J. A., Scott, P. R. & Bingham, J. (1988). Field performance of Rendezvous, a wheat cultivar with resistance to eyespot (*Pseudocercosporella herpotrichoides*) derived from *Aegilops ventricosa*. *Plant Pathology,* **37**, 251–60.

Hollins, T. W., Scott, P. R. & Gregory, R. S. (1986). The relative resistance of wheat, rye and triticale to take-all caused by *Gaeumannomyces graminis*. *Plant Pathology,* **35**, 93–100.

Hollins, T. W., Scott, P. R. & Paine, J. R. (1985). Morphology, benomyl resistance and pathogenicity to wheat and rye of isolates of *Pseudocercosporella herpotrichoides*. *Plant Pathology,* **34**, 369–79.

Huang, Jeng-sheng (1986). Ultrastructure of bacterial penetration in plants. *Annual Review of Phytopathology,* **24**, 141–57.

Ivens, G. W. (ed.) (1989). *The UK Pesticide Guide*. CAB and BCPC publication.

Jackson, G. V. H. & Wheeler, B. E. J. (1974). Perennation of *Sphaerotheca mors-uvae* as cleistocarps. *Transactions of the British Mycological Society*, **62**, 73–87.

Jeger, M. J., Jones, D. G. & Griffiths, E. (1981). Disease progress of non-specialised fungal pathogens in intra-specific mixed stands of cereal cultivars. II. Field experiments. *Annals of Applied Biology*, **98**, 199–210.

Johnson, R. (1981). Durable disease resistance. In *Strategies for the Control of Cereal Disease*, ed. J. F. Jenkyn & R. T. Plumb, pp. 55–63. Oxford: Blackwell.

Johnston, A. & Booth, C. (ed.) (1983). *Plant Pathologist's Pocketbook*, pp. 4–6. Slough: Commonwealth Agricultural Bureaux.

Kennedy, B. W. & Alcorn, S. M. (1980). Estimates of US crop losses to protocaryote plant pathogens. *Plant Disease,* **64**, 674–6.

King, J. E. & Griffin, M. J. (1985). Survey of benomyl resistance in *Pseudocercosporella herpotrichoides* on winter wheat and barley in England and Wales in 1983. *Plant Pathology,* **34**, 272–83.

King, J. E. & Polley, R. W. (1976). Observations on the epidemiology and effects on grain yield of brown rust on spring barley. *Plant Pathology,* **25**, 63–73.

Kolattukudy, P. E. (1985). Enzymatic penetration of the plant cuticle by fungal pathogens. *Annual Review of Phytopathology,* **23**, 223–50.

Kreig, N. R. & Holt, J. G. (1984). *Bergey's Manual of Systematic Bacteriology*, 9th edn, Vol. 1. Baltimore: Williams & Wilkins.

Lapage, S. P., Sneath, P. H. H., Lessel, E. F., Skerman, V. B. D., Seeliger, H. P. R. & Clark, W. A. (ed.) (1975). *International Code of Nomenclature of Bacteria*. 1976 edition. Washington DC: American Society for Microbiology.

Large, E. C. (1952). The interpretation of progress curves for potato blight and other plant diseases. *Plant Pathology,* **1**, 109–17.

Large, E. C. (1958). Losses caused by potato blight in England and Wales. *Plant Pathology,* **7**, 39–48.

Lelliot, R. A. & Stead, D. E. (1987). Methods for the diagnosis of bacterial diseases of plants. In *Methods in Plant Pathology, Vol. 2*, ed. T. F. Preece, pp. 22–36. Oxford: Blackwell.

Matthews, G. A. (1979). *Pesticide Application Methods*. London: Longman.

Mayama, S., Matsuura, J., Iida, H. & Tani, T. (1982). The role of avenalumin in the resistance of oat to crown rust, *Puccinia coronata* f.sp. *avenae. Physiological Plant Pathology,* **20**, 189–99.

Ministry of Agriculture, Fisheries and Food (1976a). *Manual of Plant Growth Stages and Disease Assessment Keys*. Key No. 1.1.1. Pinner, Middlesex: MAFF Publications.

Ministry of Agriculture, Fisheries and Food (1976b). *Manual of Plant Growth Stages and Disease Assessment Keys*. Key No. 4.2.2. Pinner, Middlesex: MAFF Publications.

Ministry of Agriculture, Fisheries and Food (1984). *Control of Pests and Diseases of Oilseed Rape*. ADAS Booklet 2387 (84). Alnwick, Northumberland: MAFF Publications.

Ministry of Agriculture, Fisheries and Food (1986). *Use of Fungicides and Insecticides on Cereals*. Booklet 2257 (86). Alnwick, Northumberland: MAFF Publications.

Mulrooney, R. P. (1986). Soybean disease loss estimate for southern United States in 1984. *Plant Disease,* **70**, 893.

National Institute of Agricultural Botany (1985). *Disease Assessment Manual for Crop Variety Trials*. Cambridge: NIAB Publication.

Neergaard, P. (1979). *Seed Pathology*, Vols 1 and 2. London: Macmillan.

Nix, J. (1988). *Farm Management Pocketbook, 18th edn.* Wye College, University of London.

Ordish, G. (1952). *Untaken Harvest: Man's loss of crops from pest, weed and disease.* London: Constable.

Parry, D. W. & Pegg, G. F. (1985). Surface colonization, penetration and growth of three *Fusarium* species in lucerne. *Transactions of the British Mycological Society,* **85**, 495–500.

Polley, R. W. & Clarkson, J. D. S. (1980). Take-all severity and yield in winter wheat: relationship established using a single plant assessment method. *Plant Pathology,* **29**, 110–16.

Priestley, R. H. (1978). Detection of increased virulence in populations of wheat yellow rust. In *Plant Disease Epidemiology*, ed. P. R. Scott & A. Bainbridge, p. 64. Oxford: Blackwell.

Priestley, R. H. & Bayles, R. A. (1988). The contribution and value of resistant cultivars to disease control in cereals. In *Control of Plant Diseases: Costs and Benefits*, ed. B. C. Clifford & E. Lester, pp. 53–65. Oxford: Blackwell.

Priestley, R. H., Parry, D. W. & Knight, C. K. (1985). Yield responses from fungicide treatment of cereal, oilseed rape and perennial ryegrass trials in England and Wales. In *Fungicides for Crop Protection*, BCPC Monograph 31, ed. I. M. Smith, pp. 383–86. Thornton Heath, Surrey: BCPC publication.

Richardson, M. J., Jacks, J. & Smith, S. (1975). Assessment of loss caused by barley mildew using single tillers. *Plant Pathology,* **24**, 21–6.

Rotem, J., Bashi, E. & Kranz, J. (1983). Studies of crop loss in potato blight caused by *Phytophthora infestans. Plant Pathology,* **32**, 117–22.

Royle, D. J. (1985). Rational use of fungicides on cereals in England and Wales. In *Fungicides for Crop Protection*, BCPC Monograph 31, ed. I. M. Smith, pp. 171–80. Thornton Heath, Surrey: BCPC publication.

Royle, D. J. & Butler, D. R. (1986). Epidemiological significance of liquid water in crop canopies and its role in disease forecasting. In *Water, Fungi and Plants*, ed. P. G. Ayres & L. Boddy, pp. 139–56. Cambridge: Cambridge University Press.

Royle, D. J. & Shaw, M. W. (1988). The costs and benefits of disease forecasting in farming practice. In *Control of Plant Diseases: Costs and Benefits*, ed. B. C. Clifford & E. Lester, pp. 231–46. Oxford: Blackwell.

Schmitthenner, A. F. (1985). Problems and progress in control of *Phytophthora* root rot of soybean. *Plant Disease,* **69**, 362–8.

Schroth, M. N. & Hancock, J. G. (1985). Soil antagonists in IPM systems. In *Biological Control in Agricultural IPM Systems*, ed. M. A. Hoy & D. C. Herzog, pp. 422–3. Florida: Academic Press.

Scott, P. R. & Hollins, T. W. (1974). Effects of eyespot on the yield of winter wheat. *Annals of Applied Biology,* **78**, 269–79.

Shaw, M. W. (1987). Assessment of upward movement of rain splash using a fluorescent tracer method and its application to the epidemiology of cereal pathogens. *Plant Pathology*, **36**, 201–13.

Shephard, M. C. (1985). Fungicide behaviour in the plant-systemicity. In *Fungicides for Crop Protection*, BCPC Monograph 31, ed. I. M. Smith, pp. 99–106. Thornton Heath, Surrey: BCPC publication.

Shephard, M. C. (1987). Screening for fungicides. *Annual Review of Phytopathology*, **25**, 189–206.

Shurtleff, M. C. (ed.) (1980). *Compendium of Corn Diseases*. St Paul: The American Phytopathological Society.

Simkin, M., Nicholson, S. M. & Clare, R. W. (1985). Take-all. *Rosemaund EHF Annual Review 1985*, pp. 12–17. Hereford.

Sinclair, J. B. (ed.) (1982). *Compendium of Soybean Diseases*. St Paul: The American Phytopathological Society.

Sisler, H. D. & Ragsdale, N. N. (1984). Biochemical and cellular aspects of the antifungal action of ergosterol biosynthesis inhibitors. In *Mode of Action of Antifungal Agents*, ed. A. P. J. Trinci & J. F. Ryley, pp. 257–82. Cambridge: Cambridge University Press.

Smedegaard-Petersen, V. (1984). The role of respiration and energy generation in diseased and disease-resistant plants. In *Plant Diseases, Infection, Damage and Loss*, ed. R. K. S. Wood & G. J. Jellis, pp. 73–86. Oxford: Blackwell.

Smedegaard-Petersen, V. & Tolstrup, K. (1985). The limiting effect of disease resistance on yield. *Annual Review of Phytopathology*, **23**, 475–90.

Smith, H. G. & Hinckes, J. A. (1985). Studies on beet western yellows virus in oilseed rape (*Brassica napus* ssp. *oleifera*) and sugar beet (*Beta vulgaris*). *Annals of Applied Biology*, **107**, 473–84.

Stakman, E. C. & Harrar, J. G. (1957). *Principles of Plant Pathology*, p. 306. New York: Ronald Press.

Tait, E. J. (1987). Rationality in pesticide use and the role of forecasting. In *Rational Pesticide Use*, ed. K. J. Brent & R. K. Atkin, pp. 225–38. Cambridge: Cambridge University Press.

Talboys, P. W., Garrett, C. M. E., Ainsworth, G. C., Pegg, G. F. & Wallace, E. R. (1973). A guide to the use of terms in plant pathology. *Phytopathological Paper No. 17*. Slough: Commonwealth Agricultural Bureaux.

Tantius, P. H., Fyfe, A. M. Shaw, D. S. & Shattock, R. C. (1986). Occurrence of the A2 mating type and self-fertile isolates of *Phytophthora infestans* in England and Wales. *Plant Pathology*, **35**, 578–81.

Tarr, S. A. J. (1972). *Principles of Plant Pathology*, p. 419. London: Macmillan.

Tottman, D. R. & Broad, H. (1987). The decimal code for the growth stages of cereals, with illustrations. *Annals of Applied Biology*, **110**, 441–54.

Vanderplank, J. E. (1963). *Plant Diseases: Epidemics and Control*. New York: Academic Press.

Vanderplank J. E. (1968). *Disease Resistance in Plants*. New York: Academic Press.

Vanderplank, J. E. (1968). *Disease Resistance in Plants*. New York: p. 97. New York: Academic Press.

Wale, S. J., Robertson, K., Robinson, K. & Foster, G. (1986). Studies on the relationship of contamination of seed potato tubers with *Erwinia* spp. to blackleg incidence and large scale hot water dipping to reduce contamination. In *Aspects of Applied Biology 13, Crop Protection of Sugar Beet and Crop Protection and Quality of*

Potatoes, pp. 285–91. Wellesbourne, Warwick: Association of Applied Biologists.

Walkey, D. G. A. (1985). *Applied Plant Virology*, pp. 32–44. London: Heinemann.

Walsh, J. A. & Tomlinson, J. A. (1985). Viruses infecting winter oilseed rape (*Brassica napus* ssp. *oleifera*). *Annals of Applied Biology*, **107**, 485–95.

Williams, P. H., Aist, J. R. & Bhattacharya, P. K. (1973). Host–parasite relations in cabbage clubroot. In *Fungal Pathogenicity and the Plant's Response*, ed. R. J. W. Byrde & C. V. Cutting, pp. 141–58. London: Academic Press.

Wilson, J. (1987). Commercial implementation of forecast methods. In *Rational Pesticide Use*, ed. K. J. Brent & R. K. Atkin, pp. 333–41. Cambridge: Cambridge University Press.

Wolfe, M. S. (1985). The current status and prospects of multiline cultivars and variety mixtures for disease resistance. *Annual Review of Phytopathology*, **23**, 251–73.

Yarham, D. J. (1986). Change and decay–the sociology of cereal foot rots. In *British Crop Protection Conference Pests and Diseases–1986*, pp. 401–10. Thornton Heath, Surrey: BCPC Publication.

Yarham, D. J. & Giltrap, N. (1989). Crop diseases in a changing agriculture. *Plant Pathology* (in press).

Yarham, D. J. & Norton, J. (1981). Effects of cultivation methods on disease. Ir *Strategies for the Control of Cereal Disease*, ed. J. F. Jenkyn & R. T. Plumb, p. 159. Oxford: Blackwell.

Zadoks, J. C. & Schein, R. D. (1979). *Epidemiology and Plant Disease Management*. Oxford: Oxford University Press.

Index

Note: The names of both the principal plant diseases (with the specific host crop(s) identified in parentheses) and the causal organisms have been indexed, with cross references from one to the other. Where a disease is known by alternative names, the one most frequently used in this book has been entered. The more significant page numbers in a series are printed in bold type; numbers printed in italic type refer to Figures; inclusive page numbers followed by *passim* indicate scattered references.